PETROLEUM EXPLORATION STRATEGIES IN DEVELOPING COUNTRIES

PETROLEUM EXPLORATION STRATEGIES IN DEVELOPING COUNTRIES

PROCEEDINGS OF A UNITED NATIONS
MEETING HELD IN THE HAGUE
16-20 MARCH 1981

UNITED NATIONS
Natural Resources and Energy Division
Department of Technical Co-operation for Development

Published in co-operation with the United Nations by
Graham & Trotman Limited

Published in 1982 by

Graham & Trotman Limited
Sterling House
66 Wilton Road
London SW1V 1DE

in co-operation with the United Nations

British Library Cataloguing in Publication Data

Petroleum exploration strategies in developing countries.
1. Underdeveloped areas—Petroleum industry and trade—Congresses
2. Gas, Natural—Underdeveloped areas—Congresses
330.8'23 HD9560.1

ISBN 0-86010-346-3

The views and opinions expressed in these papers are those of the authors and do not necessarily reflect those of the United Nations.

Printed in Great Britain by Unwin Brothers Limited, Surrey

CONTENTS

INTRODUCTION

The United Nations Meeting on Petroleum Exploration Strategies in Developing Countries, which was organized by the Natural Resources and Energy Division of the United Nations Department of Technical Co-operation for Development, was held from 16 to 20 March 1981 at The Netherlands Congress Centre, The Hague.

This meeting represented an important step in the ongoing efforts of the United Nations to focus attention on the energy problems of the developing countries and to promote co-operation at all possible levels for energy development. Related efforts during the 1970s included the United Nations *ad hoc* Group of Experts on the Supply and Demand of Crude Oil and Petroleum Products (New York, 1971),[1] the United Nations Interregional Seminar on Petroleum Refining in Developing Countries (New Delhi, 1973),[2] the United Nations Meeting on Petroleum Co-operation Among Developing Countries (Geneva, 1975),[3] and the United Nations Interregional Symposium on State Petroleum Enterprises in Developing Countries (Vienna, 1978),[4] and the United Nations International Meeting on Petroleum Geology was held in Beijing in 1980.[5]

The meeting was attended by 150 participants from 49 countries and nine international organizations as indicated in the List of Participants. Invitations had been extended to all member countries of the United Nations with an indication that participants from both the public and private economic sectors would be welcome.

The programme of the meeting was divided into the following sections:

1. Oil and Gas Potential and Exploration;
2. Petroleum Exploration Technologies;
3. Financial Aspects of Petroleum Exploration;
4. Exploration and Development Agreements;

5. Strengthening National Capacities in Petroleum Exploration;
6. Bilateral and Multilateral Co-operation Programmes in Petroleum Exploration.

Sixteen speakers prepared papers on various aspects of the programme of the meeting and these appear in this volume. All but one of the speakers presented their papers at the meeting and participated in the discussions.

During the proceedings, panel discussions were organized on Financial Aspects, Petroleum Exploration and Development Agreements and Petroleum Exploration Technologies. The contribution of the speakers and panellists to the success of the meeting was invaluable.

The formal proceedings of the meeting were followed on 21 March by informative visits of participants, especially from developing countries, to the petroleum equipment yards of De Groot International Contractors B.V. at Dordrecht and Zwijndrecht and on 23 March to the Groningen gas fields of the Nederlandse Aardolie Maatschappij.

This meeting was designed to bring together specialists with diverse and often apparently opposing interests in order to promote a constructive dialogue in the crucial area of petroleum exploration. This is an area where long-term relationships are formed. Such relationships between host governments in the developing world and transnational oil corporations are often co-operative, but adversary relationships sometimes occur.

The choice of speakers was also made with this in mind — to point out basic differences in interests but also to analyse constructively possible avenues of co-operation.

While the results of such a meeting can be judged only by the participants, it is hoped that the presentation of all possible points of view may have widened the horizons of those present and that the papers included in this book convey the urgency and the potential for action.

The main rationale of the meeting was based on the results of studies by a variety of experts who had demonstrated that exploration was lagging in the developing countries despite a general recognition that a significant proportion of the world's petroleum resources are in the developing world.

During the meeting it was reconfirmed that a serious petroleum exploration gap continues to exist between developed and developing countries with most current efforts concentrated on exploration in the industrial countries, especially in North America.

Exploration statistics left much to be desired, especially with regard to the developing countries. Nevertheless, available information confirmed the lack of an intensive exploration effort in relation to either the petroleum potential or the petroleum requirements of the oil importing developing countries.

A variety of reasons were put forward as possible explanations for this exploration gap. As far as the developing countries themselves were concerned, it was noted that generally member countries of OPEC had accelerated their own efforts thanks to the availability of financial resources which could be used to purchase services of trained personnel and technology. Despite notable exceptions (e.g. Argentina, Brazil, India) most of the oil importing developing countries have been unable to accelerate exploration of their petroleum potential because of the scarcity of financial resources, trained personnel and technology. The usual alternative, that is the involvement of transnational oil corporations in exploration, has not filled the gap, largely due to the lack of confidence between the parties concerned.

Transnational oil corporations, in view of the radical changes in the world petroleum industry during the 1970s especially with regard to revisions in long-term exploration and development agreements and various nationalization measures, generally perceived political and economic conditions in the developing world less favourably than in the industrial countries. The latter countries also instituted a variety of measures which permitted higher profitability for petroleum discoveries. However, attitudes have varied among transnational oil corporations.

Developing countries for their part remained apprehensive in entering into long-term exploration and development agreements with transnational oil corporations in view of the generally unfavourable history of such contracts in the past, their lack of knowledge of the magnitude, if any, of their petroleum potential and their relative inexperience with regard to petroleum technologies and training.

Petroleum technologies have advanced to such a degree as to permit exploration in remote and difficult environments both onshore and offshore. Rapid technological advances have been made particularly with regard to seismic surveys which have reduced, but certainly not as yet eliminated, the risk factor in petroleum exploration.

Intensified exploration, particularly in the industrial countries, resulted in shortages and higher prices for equipment and trained personnel. In other words there has been serious inflation in the costs of petroleum exploration with adverse effects on similar efforts in developing countries.

It was emphasized that petroleum exploration technologies were not the monopoly of the transnational oil corporations since their ownership was diffused among a variety of independent companies. Nevertheless the transfer or absorption of such technologies in the developing world required appropriate institutional arrangements such as the establishment and strengthening of state petroleum enterprises and the training of personnel. Only strong state petroleum enterprises and/or transnational oil corporations could train and develop an adequate number of personnel needed to manage exploration programmes in a sustained manner.

Some developing countries had succeeded in establishing strong state petroleum enterprises which have been responsible for accelerated exploration programmes in their own countries. In a few cases such enterprises had even expanded their exploration activities to other developing countries either under exploration and development or technical co-operation agreements.

Most of the developing countries have remained dependent on imported technologies through the transnational oil corporations although greater emphasis has been placed on the training of local personnel. Several developing countries have benefited from both bilateral and multilateral assistance programmes which have facilitated the transfer of petroleum technologies and helped in the establishment and strengthening of local petroleum exploration institutions.

The transfer of petroleum exploration technologies was regarded as crucial to accelerate exploration programmes in the developing world. It was claimed that such a transfer had not been undertaken to any significant degree by the transnational oil corporations in their long history of exploration; the price for this transfer had been too high and had not been limited to a monetary reward; it involved actual participation in the exploitation of petroleum discoveries which might be regarded as contrary to the principles of permanent sovereignty over natural resources. If this situation were to be improved, developing countries needed to develop their own scientific and technological capacities as had already been done by some developing countries which have succeeded in acquiring petroleum exploration technologies (e.g. Argentina, India, Mexico).

For the developing countries to engage in petroleum exploration requires adequate financial support both from national and international sources. Traditionally finance for petroleum exploration in developing countries was provided by transnational oil corporations under the provisions of long-term exploration and development agreements. Because of the structure of the petroleum industry and the discovery of ample supplies, however, exploration did not extend in any significant way to a great number of the developing countries.

Higher oil prices in the 1970s could be expected to lead to an acceleration of exploration especially in developing countries because of their petroleum potential. As indicated previously, however, this has not happened. The retrenchment of transnational oil corporations has not been made up by the efforts of the developing countries themselves or by the state petroleum enterprises from the developed countries.

In oil importing developing countries, it was estimated that a doubling of exploration effort was required just to maintain current levels of oil import dependence. Financial requirements for such a doubling of effort were estimated at $8 billion per annum. A variety of ideas were put forward for consideration including the establishment of an energy affiliate of the World Bank with considerably augmented finance for loans and credits to developing countries; the establishment of an international oil company with the participation of state petroleum enterprises of developed and developing countries for exploration in the developing world and substantially increased bilateral and multilateral technical assistance programmes.

It was emphasized that all these suggestions were put forward in a spirit of additionality since it was generally envisaged that the main sources of finance would remain with the transnational oil corporations and the governments of developing countries in the foreseeable future.

It was also pointed out that it was essential to examine ways and means to assist investors through appropriate insurance schemes so as to reduce the financial consequences of the risk factor in petroleum exploration. In this respect the concept of a Petroleum Exploration Insurance Fund was put forward. Such a fund could involve (a) a contribution of $500 - 1000 million as initial capital to be paid up over a period of a few years; (b) insurance of a percentage of exploration costs, and (c) replenishment from successful exploration ventures.

Some developing countries emphasized the success of their state petroleum enterprises in sustained petroleum exploration programmes without the involvement of transnational oil corporations; others indicated that they preferred a joint approach by which transnational oil corporations participate with state petroleum enterprises.

Most of the developing countries emphasized that transnational oil corporations were essential because of their know-how and financial strength and, therefore, exploration and development agreements were of paramount importance. Such agreements provided for all technical, economic, administrative and fiscal aspects of petroleum exploration and development. Because of a lack of information and local expertise many developing countries required advisory assistance in this crucial field. Various multilateral organizations, including the United Nations Department of Technical Co-operation for Development and the Centre on Transnational Corporations as well as the Commonwealth Secretariat and the World Bank have provided such assistance. Similarly, some developing countries with experience assisted other developing countries. Nevertheless, much more needed to be done especially with regard to the collection, analysis and dissemination of up-to-date information on recent trends in the provisions of petroleum exploration and development agreements.

Particular attention was attracted during the discussions on the possible role of the World Bank in view of its expanded programme of loans and credits in petroleum exploration and its declared policy to promote private investment in developing countries.

A variety of contracts had evolved which substituted for the old concession agreements which had become unacceptable to the developing countries. Nevertheless, new names — production sharing agreements, joint ventures, service contracts — were not enough. Thorough analysis was required in each case in order to determine the equity of the various provisions in view of the particular circumstances in each case.

Despite their concentration in exploration in the industrial countries, transnational oil corporations were active in a great number of developing countries and they regarded exploration and development agreements as contracts to be respected by both parties. In view of their risk taking, as private corporations, they expected to receive commensurate rewards in the form of profits from successful ventures to provide for the losses from failures and to obtain a return on their risk capital. The level of the expected return, however, was not defined.

It was emphasized that in the great majority of cases transnational oil corporations had co-operated effectively with host developing countries even when revisions became necessary because of unforeseen circumstances.

It was pointed out that the magnitude of the petroleum potential of the developing countries could be regarded as a source of strength in their negotiations with transnational oil corporations and petroleum enterprises from developed countries. Developing countries required information and analysis not only of the provisions of petroleum exploration and development agreements but also of the basic interests and motivations of the other parties.

The importance of this subject could not be over-emphasized and it was not limited to the developing countries as evidenced by recent petroleum laws in the United Kingdom and Canada with fundamental revisions in petroleum exploration and development agreements in those countries.

Reference was made in this respect to the prevalence of considerable economic rent in petroleum with consequent claims by both the investors and the host countries. Perhaps the elimination or reduction of the risk factor in exploration through an insurance fund as mentioned previously could contribute to a better understanding between the parties involved. It was, therefore, felt that these issues required much closer consideration and further study.

C. CONSTANTINOU
Technical Secretary and
Rapporteur

1. United Nations, '*Petroleum in the 1970s*', New York, 1974, Sales No. E.74.II.A.1.
2. *Proceedings of the United Nations Interregional Seminar on Petroleum Refining in Developing Countries*, New Delhi, Indian Oil Corporation, 1973.
3. United Nations, *Petroleum Co-operation Among Developing Countries*, New York, 1977, Sales No. E.77.II.A.3.
4. United Nations, *State Petroleum Enterprises in Developing Countries* (Elmsford, N.Y., Pergamon Press, 1980).
5. United Nations, *Petroleum Geology in China* (Tulsa, Oklahoma, PennWell, 1981).

OPENING STATEMENT BY THE REPRESENTATIVE OF THE SECRETARY-GENERAL OF THE UNITED NATIONS

Bi Jilong
Under-Secretary-General, Department of Technical Co-operation for Development

Distinguished Colleagues and Friends,

First of all, let me welcome you to the United Nations Meeting on Petroleum Exploration Strategies in Developing Countries. This is an extremely important topic, and it comes at a very opportune moment. I hope that in your deliberations you will be able to come to conclusions which are helpful to the developing countries in their search for development.

Before I say a few words about the topic of this meeting, I should like to express my sincere thanks, both on behalf of the Secretary-General and on my own behalf, to the Government of The Netherlands, which is acting as host for the meeting. The United Nations greatly appreciates the facilities provided by the Government, as well as the general efforts of the country in promoting technical co-operation for development. As you know, the raison d'être of the Department is to assist the developing countries in accelerating their economic and social development through its technical co-operation programmes. In this regard, the Department welcomes the possibilities of greater co-ordination of bilateral and multilateral programmes of technical co-operation so as to maximize the contribution they can make to solving the problems of the developing countries.

Now, let me say a few words about the theme of this meeting. Of course, you will be discussing the technical issues concerning petroleum exploration strategies. I should only like to make a few general comments.

It is quite clear that the energy problem is one of the major issues faced by the community of nations. The situation is particularly grave for the developing countries. It is especially so for the developing countries which lack energy resources. It is also clear that the energy problem has been dominated by oil, and it is only recently that attention has come to be given to other elements of energy. All projections indicate that oil will remain dominant in the world energy picture at least until the end of the current century. Therefore, while the search for new and renewable sources of energy continues, for the immediate future exploration and development of petroleum resources will require redoubled efforts. We hope that this will give us breathing time to develop and apply new technologies and energy sources. The United Nations Conference on New and Renewable Sources of Energy, to be held in Nairobi in August 1981, will analyse such prospects. At the present meeting the focus is on the conventional sources of energy in general, and exploration for petroleum resources in particular. This is an extremely important subject, and I hope that in the course of your deliberations you will be able to break new ground in developing ideas which can be helpful in solving the energy problems of developing countries.

I suggest that the question of petroleum exploration strategies be considered in the larger context of the needs of the developing countries and the promotion of international co-operation. In this regard, there are many issues which have to be taken into account. I would like to mention only four which ought to be given careful consideration when you are coming up with appropriate strategies for the exploration of petroleum.

In the first place, it is important that the question of petroleum exploration strategies be linked to the larger question of the energy needs of the developing countries, which in turn have to be viewed as an integral part of national development objectives and plans. Developing countries will find it necessary to establish and strengthen medium- and long-term action programmes on energy which are comprehensive and include petroleum possibilities and prospects. Economic development is greatly dependent on the availability of energy; in fact, the absence of energy is one of the most serious obstacles to the development of developing countries. The rise in oil prices and the conditions of its supply mean that the poor countries have to spend a major portion of their foreign exchange earnings just to meet the cost of importing oil. This adversely affects their ability to allocate resources to development activities. While it is imperative for them to be able to devote resources to exploration, including petroleum exploration, this is also a highly difficult task because of lack of finance. For all these reasons, they will have to give very serious thought to their action programmes for meeting energy needs and accelerating economic and social development.

The second point which requires particular attention is the need of the poorer countries for a flow of resources. This subject, as you know, is included in the global negotiations relating to international economic co-operation for development agreed to by the General Assembly at its thirty-fourth session. Let us hope that there will be some satisfactory resolution of the problem.

The third point I should like to emphasize is that all future programmes for energy development, including exploration, must respect the principle of the full and permanent sovereignty of each country over its natural resources. Moreover, the exploration and rational exploitation of energy resources, both conventional and non-conventional, should be promoted in accordance with the national plans and priorities of each country. While developing countries may find it necessary to obtain external financial and technical resources for oil exploration, the international community must support and facilitate the effective participation of developing countries in the production, processing, distribution and marketing of this resource.

The fourth point which I should like to single out for consideration by this meeting relates to the possibilities of technical co-operation among developing countries in the exploration and exploitation of petroleum resources. Such co-operation can usefully include not only the sharing of scientific and technological processes for the development of energy resources, but also facilitation of the flow of financial resources and investment. In this connection, regional and international financial institutions can play a much greater role in the financing of energy projects than has been the case in the past. The governments of the developing countries themselves can also contribute to the improvement of the investment climate through appropriate measures to stimulate flows of foreign investment, albeit in the framework of their national plans and policies in the field of energy.

Of course, there are many other dimensions which are equally important. For example, there is the question of implementing the role of the government in the energy field in general, and in petroleum exploration in particular. A number of countries have established State petroleum enterprises which are playing an important role in the exploration and exploitation of mineral resources. It would be appropriate to discuss the kind of support measures which are required further to strengthen the role of the State petroleum enterprises, not only through availability of greater financial resources, but also through the training of technological and skilled manpower. Thought could be given also to new initiatives to bring together the advantages of various countries in financial and skilled manpower resources in a common endeavour to develop the great petroleum potential which may exist in some developing countries.

The United Nations, on its own and in collaboration with other international agencies, has been providing assistance to the developing countries to explore and exploit their energy resources. Such assistance has ranged from the assessment of the energy needs of the developing countries to the exploration of specific mineral and energy sources. Among other things, the United Nations Secretariat and the World Bank have identified an exploration gap in the developing countries. The United Nations stands ready to help the developing countries by every possible means within the constraints of the resources and opportunities available to it. Your recommendations will be of great help to the Department in developing future possibilities. At the same time, I would hope that the main focus of your deliberations will be on the measures that need to be taken at the national level and that international measures will be brought in by way of support for national actions.

With these brief comments, I thank you again and wish you every success in your deliberations.

WELCOMING ADDRESS

P. A. van Buuren
Deputy Director-General, International Co-operation, Netherlands Ministry of Foreign Affairs

Ladies and Gentlemen,

It is with great pleasure that I welcome you here in the *Congresgebouw* in The Hague, where you will attend the United Nations meeting on Petroleum Exploration Strategies in Developing Countries. I also welcome the initiative taken by the United Nations in organizing this meeting, for it is both timely and necessary.

It is necessary because developing countries are more dependent on oil for their energy requirements than the industrial countries. In view of the difficulties now confronting them and of the difficulties to be expected in the future, developing countries should be enabled to develop and utilize their own indigenous energy sources.

It is timely because at the international level discussions are in progress with regard to the creation of additional financial facilities for the development of energy resources in developing countries. The exploration and exploitation activities of these countries are hampered by lack of funds and technology.

However, to make progress in this matter, it does not suffice simply to provide developing countries with the capital and technology they require, for not only do a number of them have to contend with a growing burden of debt and a rising balance-of-payments deficit but, in addition, existing legislation and policy on investments are sometimes such as to provide little incentive for investors of either capital or technology to consider long-term co-operation. Another inhibiting factor is the conditions which developing countries sometimes wish to include in their arrangements with multinational concerns, with the result that the latter are interested only in production from the most accessible, most promising regions, and smaller oil fields are left unexploited.

To overcome these and similar problems governments of oil-importing developing countries could work together with appropriate United Nations and other institutions in formulating anticipatory long-term energy policies, in which oil would play an important role. Improvement of the over-all energy situation has to date been greatly impeded by the *ad hoc*, fragmentary, short-term policies which have prevented some developing countries from dealing with this issue in a more effective way.

The primary objective of this meeting is to supply policy-makers in developing countries with more choices for deciding on their future energy course; but it is not only for them that it is important as is shown by the participation of a great number of representatives of industrialized countries. Although the focus of the meeting is on developing countries, it is clear from the programme that the underlying principle is the realization that the energy problem is one which confronts industrial and developing countries alike, for we deal here with a worldwide demand for scarce resources. Adequate supplies, particularly for developing countries, should be the main objective of an international energy policy, which in principle must be pursued co-operatively by all the countries and groups of countries concerned. External energy policy will accordingly have to be based on awareness of the fact that measures designed to guarantee the requisite domestic supplies must be looked at in the context of shared responsibility for guaranteeing the energy supplies of the international community, especially of the non-oil-producing developing countries. I know this will not be an easy task, but the alternative free-for-all approach is clearly much less attractive.

The discussions which you will have this week will be pursued against this kind of background. The programme of the meeting contains the ingredients that can lead to a better understanding of the possibilities and alternative routes and policies that are open to us.

THE NEED FOR PETROLEUM EXPLORATION STRATEGIES IN THE DEVELOPING COUNTRIES

Vladimir Baum
Director of the Meeting

The decade of the seventies has taught us some very hard lessons regarding the world energy situation and, in particular, has stressed the plight of most developing countries. The facts are well known and do not require detailed elaboration. We are now more painfully aware than we used to be of the finite nature of hydrocarbons and of the steeply rising costs of energy supply. We are living in the dawn of the transition period from a petroleum-based world economy to an economy based on a new and different energy mix.

Despite the emergence of a plethora of studies and projections of the shape of things to come, we are still very much in the dark as to the specifics. In fact, the prognosticators and prophets are very much at odds with one another. Questions such as the size of the world's endowment in fossil fuels, the growth rate of energy demand, the pace of interfuel substitution, the economics of non-conventional oil and gas and the scope for enhanced recovery remain today wide open.

Against this background of economic and technological uncertainty, there are nevertheless some quite broad areas of agreement. Although there may be different assessments of the hydrocarbon potential of the third world countries, there seems to be no doubt among experts about the importance of this potential and the tardiness or inadequacy of the efforts to explore for it and develop it. Also, given the present and future economics of oil, there is much greater interest than heretofore in smaller oil and gas fields. I do not intend to quote figures at this stage; I hope you will be hearing interesting views on these subjects. These and many others should provide food for fruitful discussions.

While most forecasters tend to agree that the growth of petroleum demand in the advanced countries will in all probability be more or less stationary or low by historical standards, it remains nevertheless certain that in most developing countries demand for oil and gas will continue to grow over the next two or three decades, at least in step with industrialization and the switching from non-commercial to commercial energy sources. In the light of further increases in energy prices, the balance-of-payments problem of many oil-importing developing countries will become even more burdensome and they will have to examine ever more carefully their indigenous energy potential in order both to cover as much as possible of their own requirements and — one would hope — to expand the global energy resource base. Despite the conservationist stance of some petroleum-exporting countries — that is those countries that wish to preserve their petroleum reserves for the use of future generations or prevent overheating of their economies at present — they too have an intrinsic interest in learning the extent of their potential so as to formulate appropriate economic policies.

Thus, there is a very good case to be made for pushing exploration; that is to say, in other words, nobody would deny that it is 'good' to find oil. But, exploration for oil is a complex, sophisticated and often a very long-term operation which carries with it considerable risk, as it may or may not be successful. Although progress in scientific knowledge and in technology has greatly improved the statistical success ratio, large sums of money and rare and expensive expertise are required for a well conceived exploration campaign. For the policy-maker in a developing country, in particular in capital-deficient developing countries, there are some tough choices to be made and there is more than one possible way of dealing with them.

In brief, to throw some light on the possible choices, on the opportunities and obstacles which face intensified exploration for hydrocarbons in the developing countries, the United Nations has convened this meeting. On practically every issue before you, there will certainly be divergent views supported by different experiences, perceptions and philosophies.

Let me touch upon the major topics before this gathering.

First, what do we know, or believe we know, about the yet untapped potential, the prospects of finding oil and gas? The survey of energy resources produced for the 1980 World Energy Conference indicates that some 600 sedimentary basins have been identified around the globe. So far, only in slightly more than one-quarter of them has exploration led to the discovery of known hydrocarbon deposits. Two hundred and forty basins have been partially explored without — as yet — any economic deposits being found, while, to date, another 200 basins have remained untouched by the hand of the explorer. It is true that many of these unexplored basins are found in hostile environments or areas that are very difficult of access. A large proportion of these partially or totally unexplored sedimentary basins falls within the boundaries of developing countries.

In evaluating the prospects of these basins, I am convinced that we shall have our share of optimists and 'realists'. Attempts will be made to relate our knowledge or ignorance of prospects to the intensity of exploration in most developing countries. I should, therefore, expect this gathering to begin by taking a very hard look at this state of affairs and to examine its causes realistically. In so doing, it is to be hoped that the quest or search for even minor fields will not be overlooked.

Second, our knowledge of petroleum geology and technology has greatly improved over the years. Certainly, as I have said before, exploration remains a costly and risky business, but the odds can be considerably shortened if the fruits of modern science can be put to use. The meeting will, therefore, examine the state of exploration technology and the conditions under which it can be made available to developing countries. Moreover, we must delve into the ways whereby not only know-how but also 'know-why' can be transferred and effectively absorbed. Although in many parts of the third world local expertise has made impressive strides, there will still be lacunae to be filled. The learning process will take time.

However, technology alone will not do the trick. The numerous components of an exploration programme have to be properly orchestrated, managed, supervised. In parallel with a host of specialized technicians, knowledgeable and experienced managers are required to put together the whole design of an appropriate exploration process. The many options for acquiring technology and designing the exploration programme are the second topic we shall discuss.

Third, having assessed the broad geological prospects and the promise of technology, we must give thought to the price-tag, the cost of an exploration venture, bearing in mind its risky nature. A well conceived programme moves in logical, phased, consequential steps from less expensive geological and geophysical work to more expensive exploratory and development drilling. It proceeds from less costly to more costly operations, while the risk tends to diminish; but the risk remains there until the drill hits paydirt, i.e. until drilling proves the existence of an economic hydrocarbon deposit. What may be considered economic will depend on a number of considerations. Given the trends of energy prices, what might not have been deemed economic yesterday could easily be economic today, and even more so in the future. Nevertheless, the cost of this risky investment may remain prohibitively high for a poor country besieged by innumerable claims on its scarce capital resources. The policy-maker will have to decide how much capital he or she can allocate to the exploration gamble. He will then have to look abroad to find out the conditions under which he could attract risk capital or the extent to which he could hope to receive assistance from the international community. This question could well be the central issue of this meeting.

Fourth, linked to financial requirements are the problems of negotiations and agreements with foreign investors, as well as a thorough review of the existing or desired international public funding facilities. A discussion of these matters should prove helpful to those who have to make the crucial and delicate decisions as to whether and under what conditions to embark on the search for oil within their national boundaries. Due note should also be taken of the scope of co-operation among developing countries. We are in fact witnessing today an increasing number of instances where some of the better equipped developing countries undertake exploration in other third world countries. But for the foreseeable future, particularly as we move from more alluring prospects to those where the conditions are more difficult — such as offshore prospection — the main source of know-how and risk capital will have to be sought in the experienced and well-heeled companies of the industrialized world. Thus, the ventures would have to be of mutual interest while respecting fully the host country's sovereignty over its natural resources.

Fifth, since exploratory work has already been done or is in progress with varying degrees of intensity in many developing countries, they are not strangers to the problems we have mentioned.

Most developing countries recognize the need for solid institutional structures to deal with the complexities of energy policy including the search for oil and gas. In more than 80 countries national oil enterprises have been established. Some are still in a rudimentary stage, others are already well equipped and highly sophisticated. In this respect, oil-producing countries are certainly well ahead of those which are only now on the point of embarking on a more sustained drive to find oil. In many countries these tender new shoots need strengthening; these countries need badly to develop indigenous technical and managerial expertise. This is essential if they are to negotiate as equals not only with the foreign suppliers of capital and technology, but even with the international financial institutions such as the World Bank. Such national capabilities are indispensable if the countries are to ensure that contractual obligations are properly implemented, that the activities of the foreign partners are adequately monitored and their findings properly interpreted. These national structures should be the mainstay of the policy-makers when devising the country's petroleum strategy. A thorough exchange of views and experiences on the role of national petroleum enterprises and the related problem of strengthening indigenous technological capacities is the fifth question we shall discuss.

Finally, the last topic of our deliberations should be a review of both the bilateral and the multilateral sources of technical co-operation on which the developing countries can draw. Technical co-operation programmes can help the developing countries to obtain advice on the approaches to take in regard to oil exploration, what they should do to strengthen their national institutions and how they should negotiate with foreign partners. Just to give an example, in implementation of General Assembly resolution 34/201, the Department of Technical Co-operation for Development, at the request of some forty countries, is fielding missions to assess the financial requirements over the next ten to fifteen years for energy exploration (and under energy the emphasis is on petroleum). This is just one of the various attempts to provide the policy-maker in a developing country with objective information which should be of use in formulating a policy or a strategy for exploration.

A petroleum exploration policy can be devised only in the light of the specific circumstances in each individual country. It will have to be formulated on the basis of the country's socio-economic goals and its own perception of its development priorities. There cannot be a universal prescription for all. The decisions are for each country to make. However, the building blocks on which to base the formulation of a meaningful petroleum exploration strategy can be clearly defined. They should encompass the following: (a) assumptions as to the geological prospects; (b) understanding of the technological tools and capabilities required; (c) some notion about the various options available to mobilize the necessary financial resources; (d) a policy framework within which negotiations with foreign partners can be envisaged; (e) the building or strengthening of national capabilities and structures dealing with energy in general and petroleum in particular; (f) information on the kind of assistance the international community can provide.

At this meeting, I expect to hear many, often contradictory, views on each of these subjects. I believe that a thorough, open debate will be of advantage to all. Only thus shall we leave this hospitable city enriched with new insights and convinced that this type of encounter, supported by a continuing exchange of information among all those concerned, will prove fruitful. It will, I hope, give impetus to a more systematic and sustained effort to boost exploration for petroleum in the third world.

ACRONYMS AND ABBREVIATIONS

ADC	Arabian Drilling Company
AGIP	Azienda Generale Italiana Petroli
AOC	Arabian Oil Company
ARAMCO	Arabian-American Oil Company
ARGAS	Geophysical Survey Company (Saudi Arabia)
ARPEL	Association for Mutual Co-operation among State Petroleum Enterprises in Latin America
ASCOPE	ASEAN Council on Petroleum
ASEAN	Association of South East Asian Nations
bbl	barrels
BEICIP	Bureau d'Etudes Industrielle et de Coopération de l'Institut Français du Pétrole
BNOC	British National Oil Company
BP	British Petroleum
bpd	barrels per day
CEC	Commission of the European Communities
CFP	Compagnie Française des Pétroles
CIC	Crédit Industriel et Commercial
CPE	centrally planned economy
CUF	cubic feet (gas)
D and A	drilled and abandoned
DCF	discounted cash flow
Dollars ($)	US dollars
DSDP	Deep Sea Drilling Project
ECOSOC	United Nations Economic and Social Council
EEC	European Economic Community
EGPC	Egyptian General Petroleum Corporation
ENI	Ente Nazionale Idrocarburi (Italian State petroleum enterprise)
EVRD	expected value of recoverable discoveries
GECO	geophysical company, Norway
IDA	International Development Association (World Bank Group)
IEA	International Energy Agency
IEDC	International Energy Development Corporation
IFP	Institut Français du Pétrole
IIASA	International Institute for Applied Systems Analysis
IMP	Mexican Petroleum Institute
INTEVEP	Institute of Oil Technology (Venezuela)

km^2	square kilometre
Lagoven	State petroleum enterprise of Venezuela
m	metre
MCF	1000 cubic feet (gas)
MMCFD	million cubic feet per day (gas)
NRED	Natural Resources and Energy Division of Department of Technical Co-operation for Development, United Nations
OAPEC	Organization of Arab Petroleum Exporting Countries
ODA	Official Development Assistance
OECD	Organization for Economic Co-operation and Development
OFID	OPEC Fund for International Development
OILDECO	Norwegian petroleum industry development company
OLADE	Energy Organization of Latin America
OMNIS	Office Militaire National pour les Industries Stratégiques (Madagascar)
ONGC	Oil and Natural Gas Commission (India)
OPEC	Organization of Petroleum Exporting Countries
OPIC	Overseas Private Investment Corporation (US Government)
PEMEX	State petroleum enterprise of Mexico
PETROBRAS	State petroleum enterprise of Brazil
Petromin	General Petroleum and Mineral Organization (Saudi Arabia)
psi	pounds per square inch
R and D	research and development
SAGE	Special Advisory Group on Energy (UNIDO)
SNEA	Société Nationale de l'Energie Atomique (France)
SOEKOR	State petroleum enterprise of South Africa
SONATRACH	State petroleum enterprise of Algeria
SPM	State petroleum enterprise of Madagascar
TCD	Technical Co-operation for Development, Department of (United Nations)
Tcf	trillion cubic feet (gas)
TD	total depth
Techno Export	Soviet technical assistance organization
TOE	tons of oil equivalent
tonne	metric ton
TPDC	Tanzanian Petroleum Development Corporation
UNCTAD	United Nations Conference on Trade and Development
UNDP	United Nations Development Programme
UNIDO	United Nations Industrial Development Organization
USGS	United States Geological Survey
YPF	State petroleum enterprise of Argentina

PART I

Oil and Gas Potential and Exploration

1

WORLD PETROLEUM RESERVES AND RESOURCES WITH SPECIAL REFERENCE TO DEVELOPING COUNTRIES

Michel T. Halbouty
Petroleum Consultant, Houston, Texas, USA

Even though the world has begun a transition period from conventional oil and gas energy towards alternative sources of nuclear, coal, geothermal, solar, fusion and synthetic fuels, it is evident that for at least the next three to four decades the major means of world energy survival will depend on oil and natural gas supplies. Petroleum is still King!

Numbers of possibilities exist for increasing the world's supply of energy from petroleum as well as from other sources, but lead times to achieve commercial production are very long, as a result of delays in demonstrating new technology, obtaining regulatory permits, providing environmental safeguards, developing infrastructure in remote locations, constructing facilities in harsh physical environments and, finally, completing financial arrangements.

Most explorationists believe that the undiscovered oil and gas potential throughout the world is sizeable. It is thought that about as much oil and slightly more gas remain to be found as have been discovered so far. Although these kinds of assessments are subjective, there are now enough data of various kinds to make this type of estimate credible. Beliefs that a country, state, or geologic 'province' has passed its peak of oil and gas discoveries have been erroneously put forward since the first well was drilled in the world. Therefore, it is doubtful that any basin or province anywhere in the world can be classified as thoroughly evaluated. For instance, the Gulf Coast geosyncline, one of the world's largest basins, located in the south-eastern United States, is one of the most densely drilled regions in the world. The total number of wells drilled, in all categories, in the states of Texas and Louisiana alone, which are a part of this vast geosyncline, is 820 000 — an average of one well per square kilometre. This area is definitely classified as very mature. However, exploration activity in this basin is greater today than it has ever been in the past and is expected to continue into the twenty-first century.

Table 1
Approximate prospective areas of sedimentary basins

Region	Total prospective area (1000 km^2)	Onshore (1000 km^2)	Offshore (1000 km^2)
USA	8 033	6 604	1 429
Canada	4 895	3 084	1 811
Latin America	7 459	4 843	2 616
Western Europe	3 600	1 944	1 656
Eastern Europe	1 000	900	100
USSR	14 300	10 000	4 300
Middle East	3 471	2 152	1 319
S & SE Asia	8 236	3 705	4 531
China	2 383	1 787	596
Japan	570	80	490
Australia—New Zealand	6 320	4 424	1 896
Africa—Madagascar	13 028	11 725	1 303
Antarctica	906	0	906
Total	74 210	51 248	22 953

Exploration for new reserves of petroleum and natural gas has been conducted in many parts of the world, in both what are termed 'developed' and 'developing' countries, and there remain scores of geologically attractive onshore and offshore areas which need to be extensively explored to meet the soaring demand for new supplies. Let us take a look at where these promising areas are located and their geological significance.

Table 1 shows the prospective areas of sedimentary basins in the world. The onshore area is 51 248 000 km^2, the offshore area is 22 953 000 km^2, and the total is 74 201 000 km^2. It is estimated that there are approximately 600 prospective petroleum basins in the world ranging in size from small to very large. Approximately 44% are classified as small with a size range up to 100 000 km^2; 35% are in the medium category from 100 000 to 500 000 km^2; and the remaining 21% are in the large and very large category.

Although the total prospective area in the Middle East is relatively small in comparison with the United States, Canada, Latin America, the USSR, South and South East Asia, Africa, and even Australia, it contains the most oil reserves. Also, it is only moderately explored.

Figure 1 shows the exploratory status of the approximately 600 prospective petroleum basins of the world. In order to economize on map space, many of the basins have been combined, resulting in a subjective categorization. These are the principal sedimentary basins where we may expect to find oil and gas. The basins are divided into four groupings: (1) intensely explored; (2) moderately explored; (3) partially explored; (4) essentially unexplored.

Of the approximately 600 basins in the world, about 200 have had

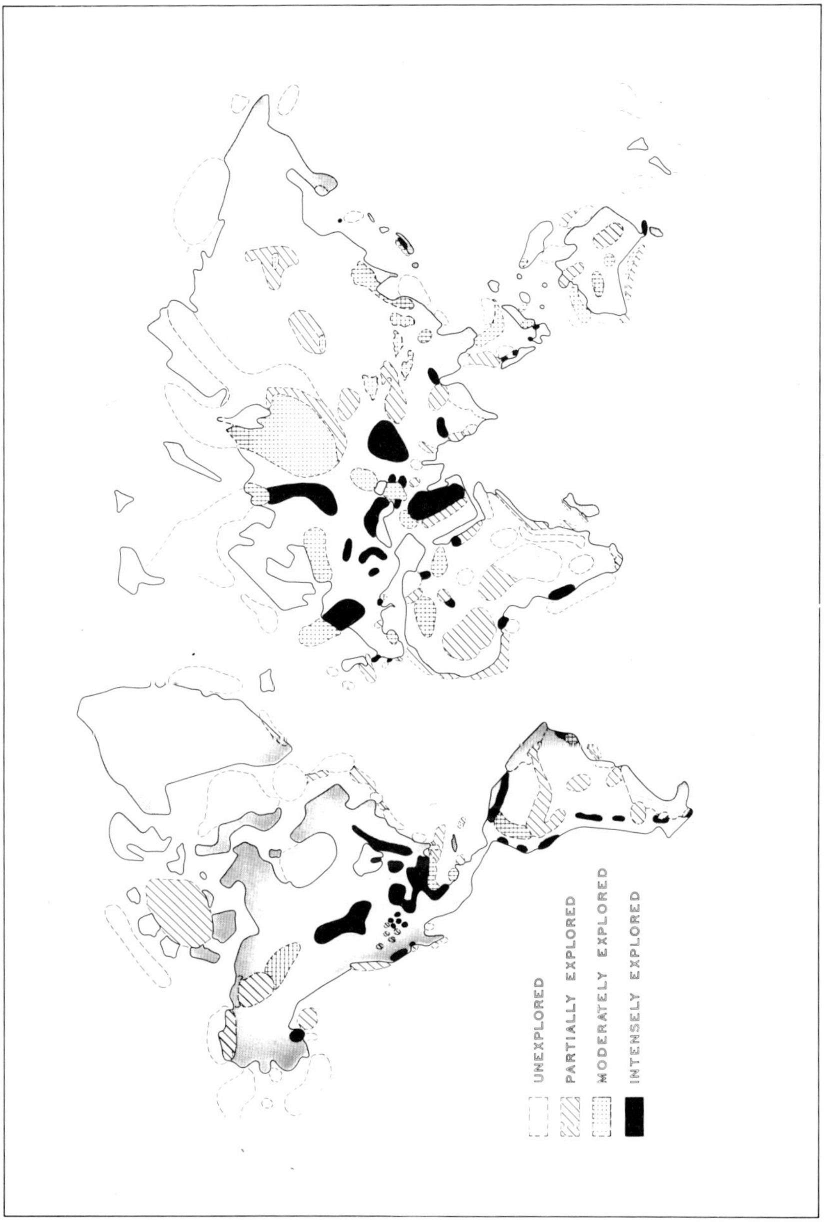

Fig. 1. Basins of the world: exploratory status.

little or no exploration for oil and gas. These are not geologic mysteries, because sufficient work has been done to indicate their dimensions, amounts and types of sediment fill, and their gross structural aspects. The vast majority of the unexplored basins are frontier areas located in harsh physical environments, that is, in the Arctic, deep water, or remote interiors of continents. The remainder are either restricted by political problems or have been bypassed because of a combination of remote locations and considerations of geologic potential.

Another 240 basins, included in the partially and moderately explored categories, have been explored to some extent, but without yielding commercial discoveries. Approximately 160 basins are productive of commercial hydrocarbons. Consequently, there are 440 known basins of the world which need to be adequately explored — and all need to be explored not only geoscientifically but by the drill.

The productive basin areas of the world, in which a significant part of the potential estimate will be discovered, are divided into two somewhat subjective categories — mature and developing (Fig. 2). It is estimated that as much as 30% of the undiscovered commercial potential oil and gas will come from mature producing areas. These are the exploratory hot spots of the past. The principal exploratory targets are: (1) extensions to known plays; (2) new plays, most probably in either stratigraphic and other subtle traps or in deeper drilling on known structures; (3) from enhanced recovery of low-grade deposits. These areas will be typified by high finding costs and a disproportionate share of both gas and small fields. The finding costs will be offset somewhat by relatively low development costs and an in-place petroleum infrastructure.

The significant petroleum frontiers of the world where sizeable accumulations might be expected are shown in Fig. 3. In these frontier areas very little exploration has been conducted. It is obvious from their locations that the hard-to-find petroleum would be in these areas. However, because the potential could be substantial, a sizeable share of our exploration efforts should be directed towards these areas. Frontier areas may be divided into moderate and harsh physical environments. Many moderate environments remain as frontier areas. Internal political processes and boundary disputes are the main reasons that the areas with moderate environments are still frontier basins.

Examples of those in which exploration is now being permitted include the United States offshore Atlantic and Pacific, offshore Argentina, offshore Brazil, the east coast of India, offshore Vietnam, and offshore China consisting of the Yellow Sea, East China Sea and South China Sea. All moderate areas are particularly significant because they are areas with potentially low finding costs and moderate development and petroleum-related costs.

Most of the harsh environment areas are in the Arctic. Difficulties in areas should be viewed as both physical and logistical in nature. They present a significant technical challenge, not just to explore but to establish production once petroleum is found. Lead times from

Fig. 2. Productive areas.

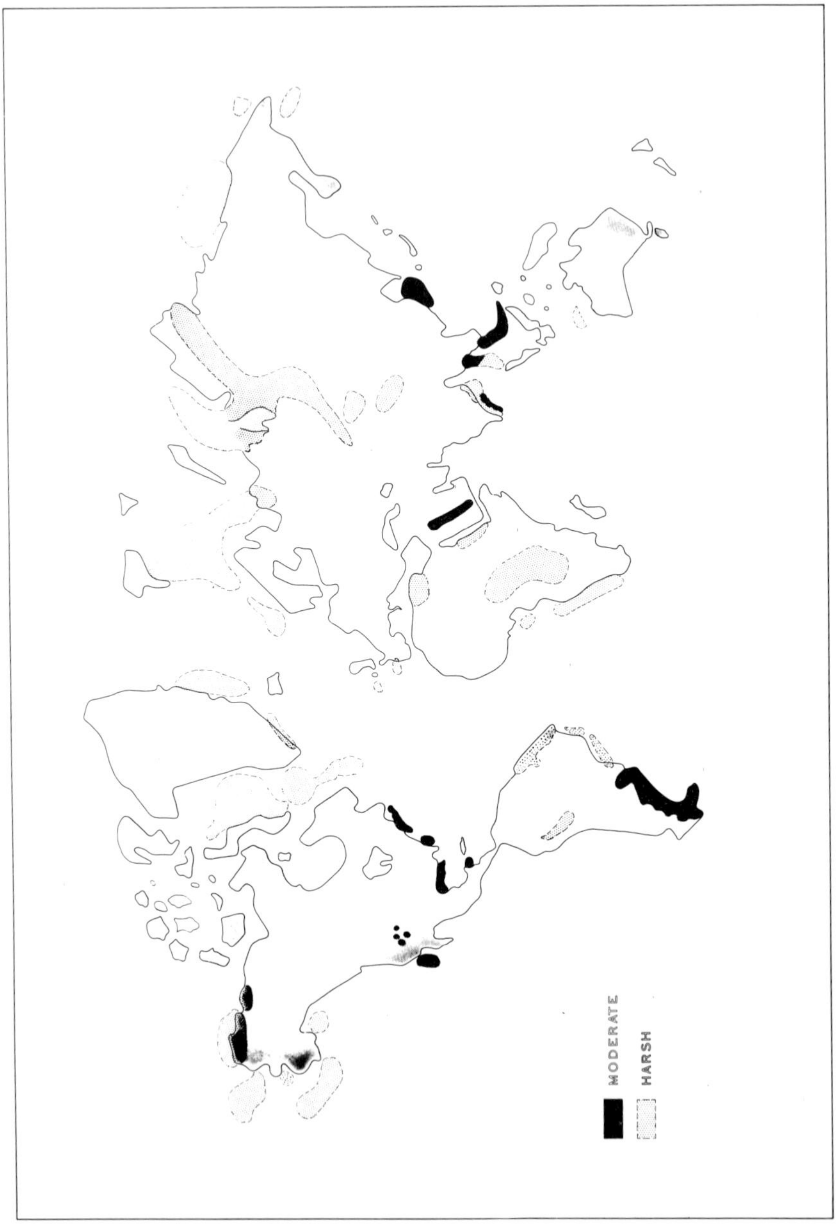

Fig. 3. Significant frontier areas.

discovery to peak production in Arctic areas could easily run to more than ten years. This time depends on weather, ice condition, water depth if offshore, and volumes of oil and gas. If government and environmental restraints are superimposed, the potential delay between discovery and production is mind-boggling.

South of the Arctic, most harsh areas are deep water — mostly in excess of 600 m. Others are in the isolated interior of the continents of Africa, Asia and South America. This potential too will be very expensive to find and to produce. Acreage availability on reasonable terms offers the real hope of accelerating discovery in the continental interiors. A full exploration cycle from acquisition, to geophysics, to drilling is what is needed in these areas. Thorough exploration procedures are required to achieve positive and conclusive results.

For example, a most exciting discovery was announced in January 1980 with the completion of the Chevron Standard Ltd Mobil Hibernia well of Newfoundland. Reports indicate that it could become the largest producer in North America. The discovery is located on the eastern offshore front which stretches approximately 1200 miles from the Canadian-United States-Maritime boundary off Maine through the Scotia shelf, the famed Grand Banks fishing grounds, the 'iceberg alley' of the Labrador coast and to the northern tip of Quebec's Ungava Peninsula. Confirmation wells have been drilled which indicate very large accumulations of hydrocarbons. The water depth in that area is 270 feet. This is a concrete example of the kind of potential that may exist in these harsh environments.

Other high potential harsh environment areas are the Labrador Sea, Baffin Bay, the Trondheim coast, offshore Greenland, and south west Ireland. The western approaches to the English Channel are in many ways similar to the North Sea, at least in type of structure and age and kind of rocks. The Kara Sea, the Khatanga trough, and the Tunguska area of the USSR are basically extensions of West Siberian geology. The Dzungarian and Tarim basins of western China also have marked similarities to West Siberia, as well as to the Kara Kum basin of southern Asiatic Russia.

The offshore North Slope and Beaufort Sea are, in most respects, comparable to their productive onshore counterparts, as are the Atlantic coast of South America, the west coast of Africa, and the deeper part of the Gulf of Mexico. The Barents Sea area off Norway and the USSR contains a variety of structures — anticlines, block faults, and salt domes. This area is productive onshore in the Penchora basin of northern European USSR. The Chukchi and Bering basins offshore between Alaska and Siberia also appear to contain numerous structures and a thick sedimentary fill.

As far as the offshore is concerned, we can be assured that technology will permit drilling to whatever water depths the operator desires. This opens up vast water areas to exploration. There are many more areas — too many to mention — located in harsh environments which have a high potential for the discovery and production of substantial quantities of oil and gas. These areas will be explored in time, and with promising results.

Frontier areas and giant oil and gas fields seem compatible. They just go together. Giants are usually defined in terms of geographic location. In the United States, Mexico, much of Canada, and Continental Europe, a field with 100 million bbl of oil or 1 Tcf of gas reserves is a giant. The same definition holds in south eastern Asia, Japan, China, Australia, central and southern Africa, and South America. However, in North Africa, the Middle East, and in Soviet Asia, the minimum size of a giant approaches 1 billion bbl or 10 Tcf. This is also true for the Arctic regions of North America. A composite compromise definition for a giant oil field is one with 500 million bbl of reserves recoverable by present methods, and a giant gas field is one which has 3 Tcf of reserves recoverable by present methods. The present petroleum giants under this definition represent less than 1% of the world's fields, but 75% of the reserves, and they account for approximately 70% of present production.

Of the giant fields, 72% are in the Eastern Hemisphere and 28% are in the Western Hemisphere. Surprisingly, these percentages are close to those of the respective land areas of the two hemispheres, excluding Antarctica, which are 69% for the Eastern Hemisphere and 31% for the Western Hemisphere. The predominance of the giants in the Eastern Hemisphere is even more impressive because of the more intensive exploration of the Western Hemisphere. It is interesting to note that, in size, approximately one-third of the world's sedimentary basins are in the USSR.

Fifty-eight per cent of the world's giant fields are within a crescent which extends from west-central Algeria through the Middle East, the southern USSR, through the Volga-Urals (Fig. 4). This area is truly the fertile crescent. In the area of the crescent, 100% of the gas and 98.5% of the oil found to date are in anticlinal traps.

The search for petroleum reserves is especially dramatic in the developing countries of the world. Every drop of oil is important today. While we must continue to explore for and develop the giant fields of the future, we must also become increasingly more aware of the many small accumulations which only a few years ago would have been termed non-commercial. These must be developed.

The developing countries of the world may be categorized as follows: (1) countries with marginal petroleum production with potential for limited export capacities, such as Peru and Egypt; (2) countries which are semi-industrialized and which are very dependent on imports of petroleum, such as Turkey and Brazil; (3) countries with extremely high balance-of-payments difficulties resulting from high import demands for petroleum coupled with scarce domestic supplies of non-commercial traditional energy sources, as in the case of Jamaica and Kenya.

Promising oil prospects are known to exist in Argentina, Benin, Cameroon, Colombia, Congo, Ghana, India, Ivory Coast, Pakistan, Philippines, Togo, Turkey, Vietnam and Zaire. And there have already

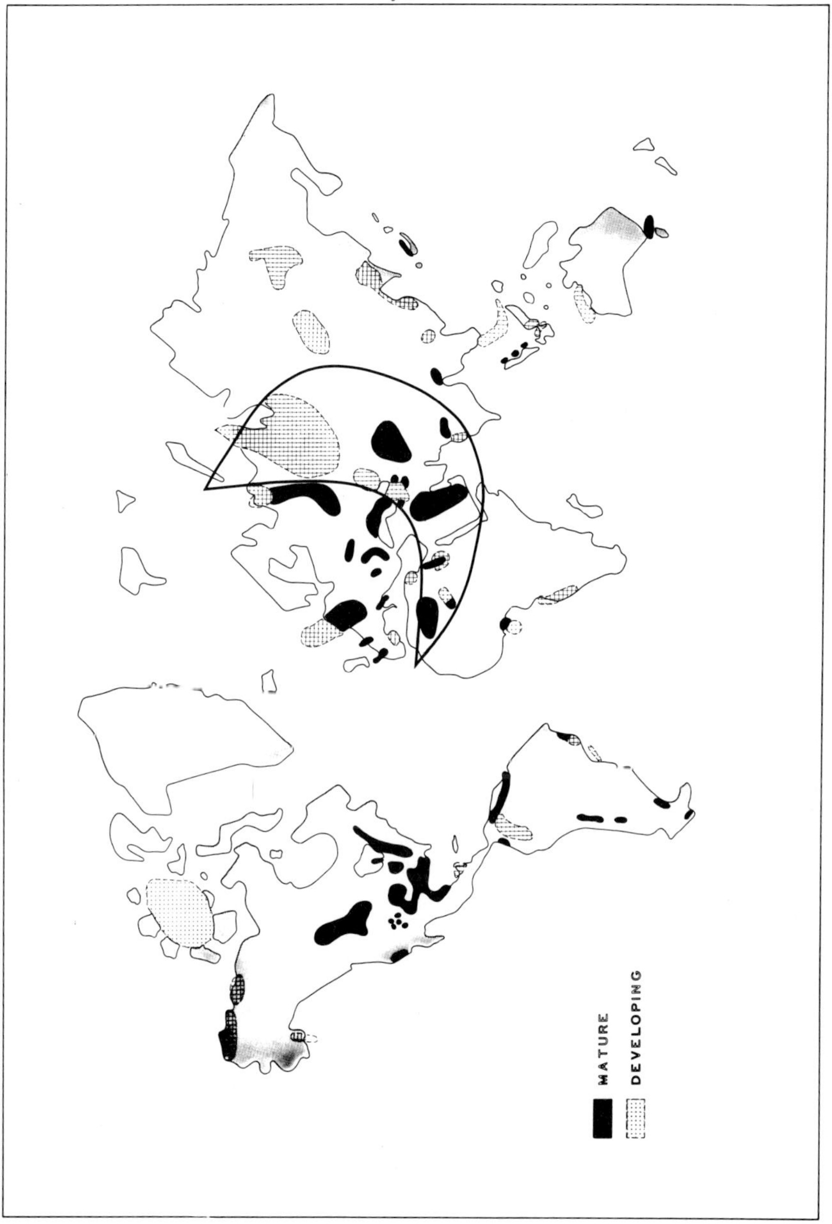

Fig. 4. Productive areas.

been discoveries of oil in five oil-importing developing countries that are not at this time petroleum producers. These are the developing countries of Benin, Chad, Niger, Ivory Coast and Sudan.

In some of the developing countries drilling activity has been increasing as of the last few years. For example, as of 1 January 1981, there are 4 drilling rigs in Angola, 8 in Cameroon, 6 in Gabon, 1 in Chad, 5 in Niger, 3 in the Ivory Coast and 2 in the Sudan — and every indication is that drilling will take place within the next few years in many of the developing countries which have petroleum potential.

An energy classification suggested by the World Bank indicates that 28 of the developing countries with a total population of approximately 1500 million are net exporters. Of course, in this classification are included the OPEC as well as non-OPEC exporting developing countries. Several, such as China, Indonesia, and Nigeria, have populations large enough to absorb most of their oil export surpluses within the next 10 years — unless discoveries of significant size are made, or unless alternative sources of energy are developed.

A large number — 92 — of the developing countries depend somewhat on imported oil supplies. Some of them, such as Chad, Ghana, Ivory Coast and Pakistan, have indications of potential to become self-sufficient within 10 years. Unfortunately, of this number, there are others, such as Brazil, the Republic of Korea and Turkey, which have an increasing demand for oil that could make them become more and more dependent on oil imports unless they develop other sources of energy. Of the 92 import-dependent countries, there are at least 64 which rely on imports for more than 75% of their national commercial supplies of energy. It is in these countries that even a small discovery of oil or a small development of other energy sources could materially reduce dependence on oil imports.

Explorationists have not yet ascertained the full petroleum potential of many of these import-dependent countries, but data have been collected and interpreted which indicate that large, easily exploitable reserves are not present. But, and most importantly, as was previously stated, even oil discoveries of modest size would greatly relieve their energy dependence burdens.

Developing countries are not necessarily under-developed scientifically or technologically, but they are just now emerging into the exploration and production phases of petroleum development. Mexico is an excellent example of a country which, after the rapidly escalating oil price hikes of the early 1970s, made the development and production of its natural resources a top national priority. The petroleum infrastructure had been established, but in 1976 Mexico's oil reserves were listed at only 7 billion bbl. Current oil reserves now approach 50 billion bbl in the proved and much more in the potential categories.

Prospects for added reserves in the developing countries include the following: (1) second-phase discoveries in older established basins; (2) better technology and exploratory methods; (3) offshore extensions of coastal basins; (4) deep drilling. Additional reserves are also possible from enhanced recovery techniques, including the recovery of heavy oil.

It is obvious that to explore, develop and produce oil and gas requires the development of sophisticated technology. However, technical developments can be either spurred on or inhibited by expectations of acreage availability, and I might add by reasonable terms for performance along with a reduced government take in successful ventures. But many governments appear to be wary of providing acreage without in-place technology, particularly regarding environmental and safety safeguards. This factor could explain why many of the developing countries show little or no reserves. It is not because there is a lack of interest in exploration, but because political or governmental conflicts have restricted the full-scale operation of their petroleum industries.

The world today has become so interconnected that the development of all countries must be considered jointly. The need for critical supplies of energy is most desperate in many of the developing countries — to sustain their irrigation projects — to provide adequate fertilizers — to fuel their industrial complexes. It is noteworthy that Indonesia, with its clean, valuable oil products, has escaped an otherwise gloomy outlook — and has an excellent opportunity to improve the lives of more than 100 million people with the further development of its total petroleum potential.

Developing nations have approximately 75% of the world's population — and produce only about 22% of the world's economic output. It is estimated that for the rest of the century, commercial use of energy in the developing nations may be expected to rise 2–4% faster than in the industrialized countries.

It is of interest to note that in the third quarter of the twentieth century, the developed nations of the world doubled their energy consumption while the developing nations quadrupled theirs. In terms of per capita consumption, that is equivalent to a 1.7 increase for the advanced countries, and 3 times as much for the developing nations. This trend will probably continue; as the more industrialized countries travel the energy road from the use of fossil fuels towards alternative sources, the developing countries will need more and more conventional energy to keep their emerging industries and agricultural programmes viable.

It should be emphasized very strongly that it is vital for the developed countries of the world to encourage and assist the developing countries to develop and produce all of their energy and natural resources to the fullest extent, as it is of paramount importance to the remainder of the world. As the developing countries produce more of their own needed resources, the amount of financial aid which other countries provide can be reduced in direct proportion.

According to World Bank data, for the remainder of this decade at least 35–40% of the world's commercial energy demands will be met by petroleum, as will a much higher proportion of the demand for commercial energy in the oil-importing developing countries.

Estimates of world proven oil and gas reserves are shown in Tables

2 and 3. The world can be divided into three major socio-economic categories: (1) the developed market economies; (2) the centrally planned economies; (3) the developing countries.

Estimates vary from source to source, just as access to information varies in different parts of the world. Through the years, I have observed that reserve estimates of oil and gas are based not only on the appraiser's knowledge but also on his optimism. Assuming that the knowledge of the appraisers is the same, it is the degree of optimism which controls the final results. Such a procedure can be referred to as the 'principle of minimal astonishment'. For my part, I am an optimist, and I foresee

Table 2
Estimated world proven oil reserves (billion barrels)

	DeGolyer[a]	World Bank[b]	Halbouty[c]	Average
Developed market economy countries	58	59	69	62
Centrally planned economy countries	88	90	104	94
Developing countries (net importers + net exporters)	480	492	570	514
Total world reserves	626	641	743	670

a DeGolyer, 1980.
b World Bank, 1980.
c Michel T. Halbouty, independent research, 1981.

Table 3
Estimated world proven natural gas reserves (trillion cubic feet)

	DeGolyer[a]	World Bank[b]	Halbouty[c]	Average
Developed market economy countries	457	497	508	487
Centrally planned economy countries	939	1021	1043	1001
Developing countries (net importers + net exporters)	1142	1242	1269	1218
Total world reserves	2538	2760	2820	2727

a DeGolyer, 1980.
b World Bank, 1980.
c Michel T. Halbouty, independent research, 1981.

billions of barrels and trillions of cubic feet of oil and gas still to be found in basins that have not yet even been adequately explored.

Analysis of representative data shows an average world total of proven oil reserves of 670 billion bbl. Of this total, 9.4% of the reserves are located in the developed market economy countries, 14% in the centrally planned economy countries; and 76.7% in the developing countries. The distribution worldwide of natural gas reserves is quite different from the distribution of oil. Of the average estimated world proven natural gas reserves of 2727 trillion cubic feet, 18% is located in the developed market economy countries, 37% in the centrally planned economy countries, and 45% in the developing countries.

My estimate of the ultimate recoverable crude oil supplies of the world is 2275 billion bbl. This includes 366 billion bbl produced through 1979, 743 billion bbl of estimated remaining reserves, and 1166 billion bbl of potential future recoveries.

My estimate of the ultimate recovery of the world's natural gas supplies is 7900 Tcf. This includes 824 Tcf already produced through 1979, 2820 Tcf estimated remaining reserves, and 4256 Tcf of potential future recoveries.

It is my thesis, first, that worldwide there will be found as much oil in the future as has been found in the past and more gas — because, geologically, the potential is out there in the unexplored low lands, high lands, swamps, ice, deserts, and forests, and under the waters of the seas; and, secondly, that the need to find the oil and gas is critical. Those two factors add up to continuous exploration, much of which will be centred in those developing countries that will most benefit from this search. As long as there is a petroleum potential anywhere in the world, and especially in the developing countries, the hunt to discover and produce will go on and on and on. This is a challenge which cannot be ignored — a challenge that faces *all* of the countries, but most of all the thrust of the challenge is hurled at the developed countries to get the job done, but this can only be accomplished with the full co-operation of all of the developing countries.

It is extremely urgent to search for and find new petroleum reserves in order to establish enough lead time for the world's scientific and technological community to research, develop and produce sufficient quantities of the other sources of energy that are needed if long-term energy requirements are to be met.

The future of the world's petroleum industry depends to a great extent on political developments. Governments, politicians and exploration companies must agree on proper regulatory frameworks which will enhance exploration for and development and production of the needed new petroleum reserves. Only proper economic and political frameworks will provide the incentives needed for these activities.

If the wildcatters are going to meet the challenge to find the oil and gas that are so desperately needed, we must begin a sharp turnaround in both exploration activities and results. The responsibility for this falls equally on private companies, government companies and all governments — irrespective of their political ideologies. This responsibility can be shared equitably through co-ordinated and effective joint ventures among the parties involved.

One key observation is most noteworthy. The developing countries — which also include *all* of the OPEC group — cannot find their own ultimate future reserve discoveries alone — for lack of know-how and technological and scientific background. And the world's people cannot wait for them to catch up on technology and science. They will need and should have the full assistance of the private sector. To be more specific, they need the multinational companies, the private sector, to discover, develop and produce their future reserves and potential. National pride should be cast aside for the realities which face the world.

There is absolutely no question in my mind that unified co-operative exchange and alignment of governments and private companies with respect to science, technology, know-how and experience will be more effective and beneficial for all concerned — than what we have seen hitherto — especially for the people of the world. What I am emphasizing is that there must be co-ordinated and effective joint ventures among governments, government companies and private companies in the exploration, development and production effort for the world's new petroleum supplies of the future — and, by this, I mean a true partnership. Without this true partnership our estimates of future producible reserves become inconsequential and meaningless.

BIBLIOGRAPHY

American Petroleum Institute, *Petroleum Data Book*. Petroleum Engineer, Dallas, 1947-.

DeGolyer, E.L. and McNaughton, L.W., *Twentieth Century Petroleum Statistics 1980*. DeGolyer & McNaughton, Dallas, 1980.

Halbouty, M.T., 'Future Trends in Exploration to Discover New Hydrocarbon Reserves Throughout the World and the Role to be Played by OPEC Member Countries' National Oil Companies', *Proceedings of the OPEC Seminar on the Present and Future Role of the National Oil Companies* (Vienna, 10–12 October 1977), pp. 14-47.

Halbouty, M.T., 'Methods Used, and Experience Gained, in Exploration for New Oil and Gas Fields in Highly Explored (Mature) Areas', *Bulletin*, AAPG, Vol. 64, no. 8, 1980, pp. 1210-1222.

Halbouty, M.T. and Moody, J.D., 'World Ultimate Reserves of Crude Oil', *Proceedings of the 10th World Petroleum Congress*, Vol 2. Heyden, London, 1980, pp. 291-301.

Potter, N., 'Hunt the Thimbleful', *Petroleum Review*, May 1980, pp. 9-10.

Safer, A., *International Oil Policy*. D.C. Heath, Lexington, Va., 1980, pp. 101-103.

Teller, E., *Energy from Heaven and Earth*. W.H. Freeman, San Francisco, 1979, 322 pp.

World Bank, *Energy in the Developing Countries*, August 1980, 92 pp.

2

OIL AND GAS POTENTIAL IN DEVELOPING COUNTRIES: PROSPECTS FOR AND PROBLEMS OF ITS DEVELOPMENT

Peter R. Odell

Director, Centre for International Energy Studies, Erasmus University, The Netherlands

INTRODUCTION

The prospects for a stable or even a declining world-wide use of oil in the immediate future and even over much of the rest of the century enable one to hypothesize about the presence or absence of a physical oil resource problem as the main determinant of the evolving pattern of world oil supply. Contemporary volumes of proven oil reserves may confidently be expected to appreciate in size as they have done over the whole history to date of the oil industry. Theoretically, therefore, such proven reserves are sufficient to cope with oil demand for as far into the future as one needs to plan.

There is, nevertheless, a real short- to medium-term problem over the supply of oil but this must be defined somewhat differently. It emerges from the acceptability and the accessibility of those oil reserves which have already been proven and which, as a result of a series of historical accidents arising largely out of recent imperial and colonial history, are heavily concentrated in a relatively small number of countries. These are mainly Middle Eastern countries in which, for reasons related essentially to the political relationships of the Great Powers in the nineteenth century and the early part of the twentieth century, the international oil companies were, indeed, strongly motivated to concentrate their attention. The development of the oil resources of the Middle East, in other words, was not only highly favoured by the region's geological characteristics as a habitat of oil; it has also emerged from the fact that the Middle East was the 'happy hunting ground' for the Anglo-American oil companies, which, until very recently, were able to secure conditions there for the exploitation of oil which were unachievable throughout most of the rest of the world. Indeed, elsewhere in the world, factors of nationalism and of anti-colonialism succeeded in inhibiting the freedom of the international oil companies to pursue their searches for low-cost oil resources. This has been most notably the case in Latin America, where the nationalization of the resources and the oil industry dates back many decades in some countries, but it has also been an important consideration in extensive parts of Africa and South East Asia.

Thus, 'accidents' of political and economic history have helped to determine the contemporary geographical pattern of the world's proven reserves of oil. Such accidents have also been responsible for the uncertain availability of those reserves, given their concentration in politically unstable parts of the world. Indeed, the present uncertainties which influence their availability are so high that the more than 500×10^9 barrels of proven and probable reserves from the countries concerned cannot provide a guaranteed continuing and adequate flow of oil. This is the case even though the volume of oil which is now required from these areas is less than 10×10^9 barrels a year and even though the amount required each year now seems likely to be a declining one. Over the rest of the century the amount needed in total will amount to a maximum of 200×10^9 barrels, less than 40% of currently proven reserves.

The extreme uncertainty over the availability of oil from the theoretically much-more-than-adequate proven reserves of the Middle East constitutes the heart of the short-term oil supply problem. This uncertainty need not, however, constitute a problem beyond the mid-1980s because alternative oil reserves which are capable of sustaining the production of the supplies needed over and above those which the Middle East countries are prepared to supply can be developed and could be produced — given appropriate national and international policies towards the oil industry.

ALTERNATIVE SUPPLIES

A survey of all the possibilities for the geographical diversification of reserves and of production potential is clearly impossible in this paper. There are, however, a number of important general points which can be made concerning such potential diversifications.

Conventional oil supplies

The areas of the world with the proven reserves which dominate the contemporary world proven reserves picture, except in the centrally planned economies, are a very small part indeed of the total potential regions for oil and gas occurrence. This can be seen in Fig. 1, in which the world's potentially petroliferous regions (including those on the continental slopes as well as on land and on the continental shelves but excluding Antarctica and the deep oceans) are defined as well as the scale of the map and the contrasting regional availability of data will allow.

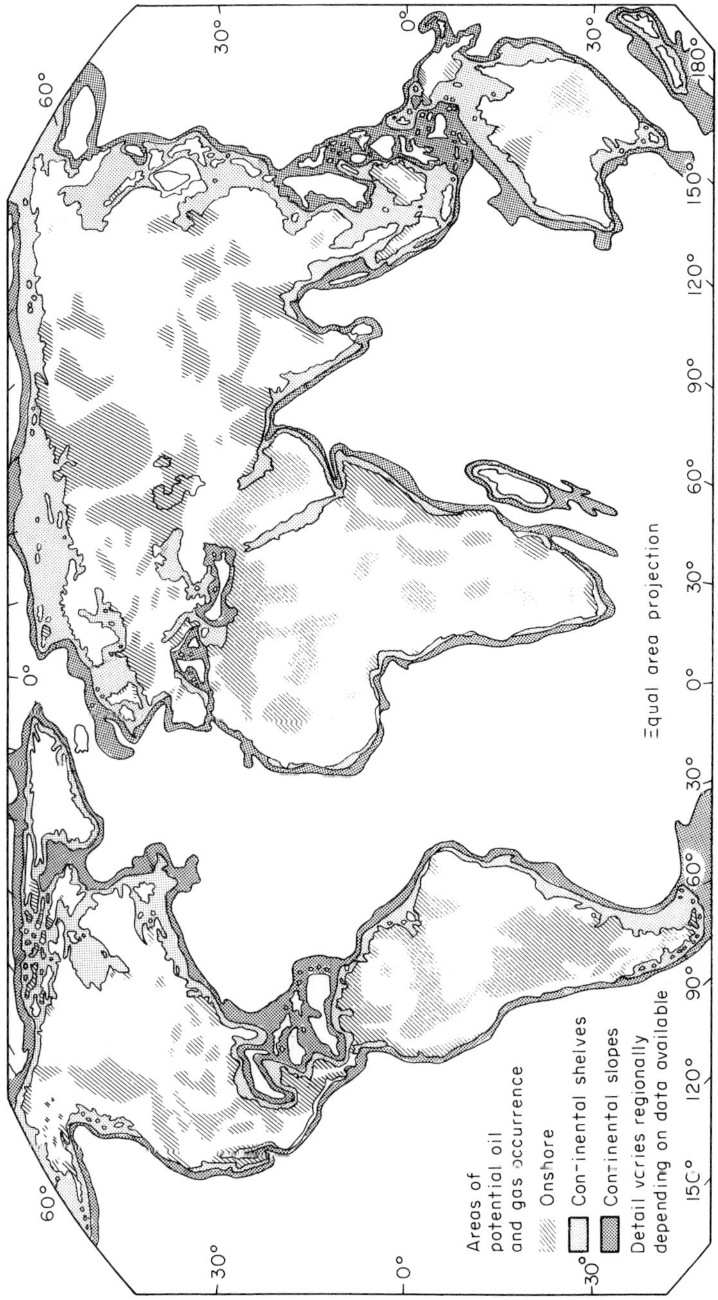

Fig. 1. The world's potentially petroliferous regions.

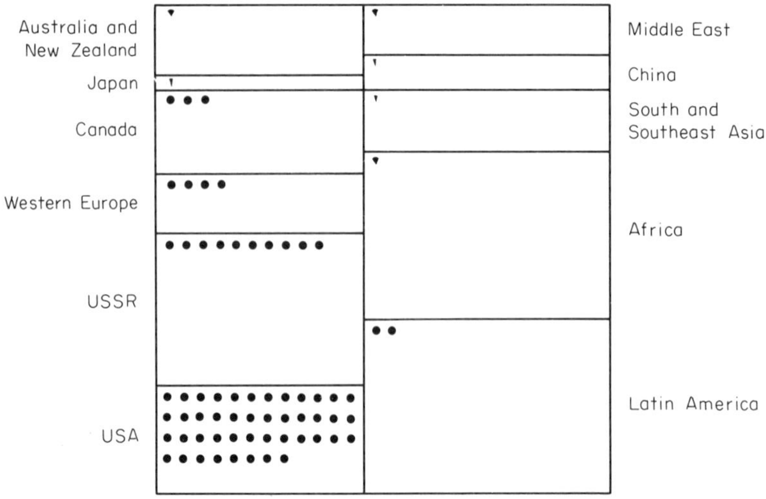

Fig. 2. The regional distribution of the world's petroliferous areas in proportion to the world total. Within each region the number of exploration and development wells which have been drilled is shown: each full circle represents 50 000 wells. Relative to the United States all other parts of the world — but especially the regions of the third world — are little drilled for their oil. Source: Grossling (1976).

From Fig. 2, however, we can see just how little oil exploration and development effort (as measured by the number of wells drilled in the different world regions) there has been in most parts of the potentially petroliferous world. This suggests that the industry has not yet progressed much beyond its infancy in terms of an effective worldwide search for opportunities for oil and gas occurrence.

Even in the developed world (on the left hand side of the diagram in Fig. 2) most regions of petroliferous potential have so far seen little of petroleum development compared with the United States. There, however, in spite of the long and intensive history of the exploration and development efforts by many hundreds of oil companies, it is still generally agreed that there are considerable quantities of new oil yet to be found in the United States. It is, indeed, this belief which forms the basis of the United States' oil pricing policy. Without the frequently expressed convictions of the country's oil companies that higher prices for domestic oil will stimulate the search for and the discovery of new reserves, the price of United States oil would have remained regulated. Given these expectations for new indigenous oil reserves of significance to the United States (in relation, that is, to its demand for some 18 million bpd of the commodity), it seems entirely inappropriate to exclude the likelihood of additional large reserves of oil (relative to levels of local use) in other parts of the developed world. What is necessary in these regions to ensure that that potential is achieved is the establishment of an acceptable *modus vivendi* between the governments concerned and the oil companies which are capable of undertaking the developments.

More particularly, however, one notes (from the right hand side of the diagram in Fig. 2) that the third world of Latin America, Africa and South East Asia contains almost 50% of the world's total potentially petroliferous regions (excluding Antarctica and the deep oceans). By contrast, however, petroleum development efforts to date in these regions have been minimal. They have amounted — and, moreover, they continue to amount — to less than 5% of the total worldwide effort. This absence of adequate attention to the potential oil and gas resources of most of the third world appears to constitute the most important failure in the development to date of the world oil industry. It also constitutes the most important challenge for the industry in the future. I shall return to discuss the issues involved later in this paper.

Unconventional oil resources

The view taken of oil resources so far, however, relates only to so-called conventional oil and thus excludes all of the world's unconventional oil resources. None of these has yet been tapped to any more than a minor — indeed, an experimental — degree. The oil shales and tar sands of North America, the shales of Brazil, Madagascar and India and the heavy oil of the Orinoco belt of Venezuela constitute just part of this enormous additional potential oil resource base. The quantities of these additional resources of oil in place are, in round terms, an order of magnitude greater than the world's present proven reserves of conventional oil. The high costs and the great difficulties which are involved in developing them are, however, well known. It thus seems prudent largely to discount their significance until good evidence is available to show that the technological and environmental problems associated with their exploitation on a large scale can be overcome. In doing this, however, let us not forget that the non-conventional oil resources do provide a hydrocarbon reserves base of immense dimensions for utilization in the twenty-first century. Indeed, together with the even larger unconventional resources of natural gas — such as geopressure gas, gas in tight formations and gas hydrates in the Arctic and under the oceans — when exploited they could ensure the continuation of our economic and societal systems based on the use of oil and gas in the twenty-first century, if, in the meantime, no acceptable forms of infinite energy sources (including coal) prove to be exploitable on a large scale — or only exploitable at a supply price above that of the non-conventional oil and gas resources.

THE POTENTIAL OF THE DEVELOPING COUNTRIES

Let us return, however, to our consideration of the nearer term future in respect of the energy outlook for the world outside the centrally planned economies for the period between the mid-1980s and the end of the first quarter of the twenty-first century. For this time period the world's most significant energy resources seem likely to be those relating to the oil (and the natural gas) potential of the many unexplored, petroliferous regions of the third world. Unfortunately, the potential opportunity for oil developments in these regions is not geographically coincident with the areas in which the international oil industry is mainly interested. On the one hand, the member companies of that industry have, over the past few decades, become unacceptable in most countries of the third world. On the other hand, the companies themselves have largely written off those potentially petroliferous parts of the world as regions in which they are prepared to risk their investments. The consequences of this serious geographical mismatch are twofold.

First, by virtue of their virtual exclusion from such large parts of the world, the international oil companies have written down, or have even written off, most of the developing countries as potential oil producers. The companies have, indeed, even generally failed to update their evaluations of the potential of most third world petroliferous regions in the light of changing theories and hypotheses on the patterns of oil and gas occurrence and of rapidly developing geological knowledge. This is clearly demonstrated in Table 1, in which the companies' outdated ideas of the third world as a habitat for recoverable oil are compared with recent figures produced by other entities with a high and continuing interest in the hydrocarbon resources of the developing countries.

Table 1
Estimates of the ultimate oil resources of the third world (in bbl x 10^9)

	Oil industry views [a]	Grossling (USGS)[b]	Min. of Geology USSR[c]
Latin America	150—230	490—1225	620
Africa	120—170	470—1200	730
South SE Asia	55—80	130—325	660
Totals	325—480	1090—2750	2010

(a) Based on figures in R. Nehring, *Giant Oil Fields and World Oil Resources*, Rand, Santa Monica, June 1978. Table on p. 88 adjusted to give comparable geographical coverage with (b) and (c). Nehring also states (p. 88) that his figures are 'roughly similar' for these regions to those published elsewhere by oil industry observers.

(b) B. F. Grossling (United States Geological Survey), 'In Search of a Probabilistic Model of Petroleum Resources Assessment', in *Energy Resources*, M. Grenon (Ed.), International Institute for Applied Systems Analysis, 1976.

(c) Visotsky *et al.*, Ministry of Geology, Moscow, 'Petroleum Potential of the Sedimentary Basins in the Developing Countries', *ibid.*

Note, for example, how inappropriate the oil industry's views on the potential resources of Latin America look in the light of recent events in Mexico. That one country's proven, probable and possible reserves are, in themselves, now approaching the oil industry's ultimate resources figure for the whole of the continent, and Mexico's ultimate resources possibly exceed the Latin American total figure by a factor of two to three.

Such an understatement of the third world's potential is, of course, quite justifiable from the companies' point of view. To them the effective resource base is the one which they expect to be able to exploit. If their relations with most countries of the world are currently so bad that they cannot conceive of circumstances in which they could work in them, then the companies must discount the resources concerned. However, the international oil companies are institutions which are changeable, and/or substitutable and/or capable of being complemented by other sorts of oil institutions. The fact that we currently have a set of institutions which, as a consequence of their history and present nature, cannot find and develop most of the world's oil resources does not mean that those resources can never be exploited. In other words, the fact that the oil companies, for their own internally justifiable reasons, are severely underestimating the oil resources of large parts of the world is no reason for the resources simply to be declared non-existent and so largely to exclude the possibility of an expanded and diversified oil supply from such countries for use later in the century. By then, revised or entirely new institutional arrangements for meeting the challenges of the oil and gas opportunities offered by the currently under-explored countries of the third world can be created.

Second, apart from this information and interpretation defect in our present appreciation of the oil resource situation in much of the potentially petroliferous world, there is a second, and, arguably, an even more important negative aspect to the developmental prospects for creating a geographically more diversified oil reserves position. This is related to the pattern of availability of the management and technological expertise and of the financial resources which are required for major oil discoveries and developments. Ninety per cent or so of the financial and human resources exist within, or are exclusively available to, the international oil companies and a small number of other entities — especially in the United States and Western Europe. As a consequence the possibilities for the exploitation of the oil and gas resources of the third world are severely limited. Local State oil companies, even when supported by State oil entities from other parts of the world, are not usually an adequate substitute, especially in managerial and technological terms, for the private sector oil companies. This has little or nothing to do with public versus private sector principles. It is more basically related to the contrast in size and in the length and depth of experience between the two sorts of entities.

THE IMPLICATIONS ARISING FROM THIS SITUATION

This institutionalized dichotomy between those world regions with prospects for large-scale oil developments, on the one hand, and, on the other, the non-availability to these regions of the resources required to develop them appears to constitute the main barrier to the rapid geographical diversification of world oil reserves and the development of an enhanced supply potential from hitherto un- or under-exploited countries. The continuation of the present situation means, moreover, that scarce capital and managerial resources for oil and gas exploitation will inevitably be used in relatively low-rates-of-return ventures in 'safe' countries, rather than in finding and in producing the world's lowest cost resources in parts of the third world. A highly relevant and important example of the way in which this geographical misallocation of the factors of production works can be seen in the contrast between the ready availability of financial and other resources for the near-future development of the high-cost oil and gas resources of the United States (where expertise and money will be absorbed in projects to win oil from shales and from coal or in working over, for the *n*th time, petroliferous regions that have already been thoroughly exploited over a period of many decades simply in the hope of finding marginal extra quantities of producible oil) and, on the other hand, the lack of adequate resources — both managerial and financial — for developing the low-cost oil of some of the massive new fields in southern Mexico. There, development costs for the large scale production of oil appear to be of the order of $2 per barrel and the number of potential opportunities in that country currently appear to be limitless, compared with the resources available to develop them.

Indeed, the development of southern Mexico as a possible second Saudi Arabia or even a second Middle East — if offshore, Central American and adjacent Caribbean potentials are also taken into account — emphasizes the sorts of opportunities for oil capital and management which still exist in the world. At the same time, of course, it also illustrates the factors which inhibit their use; factors, that is, which relate, in the final analysis, to the history of the international oil companies in a country which is still concerned for its effective national independence. It is highly unlikely that Mexico will prove to be the only major prospect in the so far undeveloped world of oil which could provide an element of highly desirable and very necessary geographical diversification of reserves discovery and development and of production potential. In the context of the world's (excluding the centrally planned economies) relatively limited needs for oil over the next fifty years, such diversification could provide the means for ensuring the well-proven use of hydrocarbons as an effective basis for economic development.

THE GEOGRAPHICAL DIVERSIFICATION OF RESERVES: DEVELOPMENT AND SUPPLY POTENTIAL

It is reasonable to infer that great potential benefits would flow from an approach to the future supply of oil and gas which was based on the geographical diversification of the oil industry's activities. All potentially oil-rich countries have a clear economic motivation, in a world of high-price energy, to produce any oil which is found in their own national territories. Thus, one may assume that each new producing country will try to add as quickly as possible to the over-all geographical diversity of supply. Such developments will not only help to bring the worldwide strategic politico-economic problems associated with undue dependence on a limited number of OPEC producers under control, but they will also help to reduce the expected continued upward pressure on oil prices. They will also serve to reduce the political pressures on the OPEC member countries in respect of their existing too-heavy burden of responsibility for sustaining the international economic order.

Recognition of the validity of this approach to the future pattern of oil developments has already emerged. President Carter referred to it in his Energy Address to the Nation in 1979 when he commented, 'there are several potentially abundant . . . sources of foreign oil and gas that have not yet been fully explored or developed . . . The Administration is developing a broader international strategy for increasing . . . exploration in these areas'. Unhappily there has been little evidence of that strategy being implemented over the last two years. Much more encouragingly, the World Bank has initiated appropriate action. It launched a modest programme for investments in oil production in third world countries in 1979 involving the commitment of some \$450 million over five years to a programme designed to accelerate petroleum production in the developing countries. That programme was small in relation to the opportunities and the needs but it did nevertheless represent a major breakthrough in World Bank activities. For many years the Bank refused to invest in oil-related activities at all. More recently, in late 1980, following the publication of its paper, *Energy in the Developing Countries* (Washington, August 1980), the World Bank has indicated its recognition of the opportunities for annual investments of billions of dollars in energy (especially oil) developments and the first steps towards the creation of a subsidiary of the Bank, an International Energy Corporation, have already been taken. This could also be the means whereby private risk capital was attracted to the ventures sponsored and/or partly financed by the World Bank. Some of the major oil companies — including Exxon, Gulf and Shell — have, indeed, already let it be known that they see good opportunities for productive investment in oil developments in third world countries and that they would welcome suggestions as to how they might become involved in a way which would minimize both the political problems and the high investment risks associated with such projects. Finally, an increasing number of third world countries now appear to be more willing to reconsider their previous hostility to, and often their rejection of, the involvement of the international oil companies in their oil sectors, provided they can be assured of the effective depoliticization of the companies involved in such activities. The new relationships which have been developed in recent years between the oil companies and the members of OPEC — particularly in respect of the managerial and technical agreements and the companies' willingness to work for a flat per-barrel fee in the context of such agreements — offer one possible approach to this difficulty.

In investment terms what appears to be needed for energy production developments in the third world is an availability of up to $8000 million per year, most of which for at least a decade would go into the oil and gas sectors. This investment, which is needed to pay for much of the necessary managerial and technical expertise, as well as for the hardware and the services involved, would make it possible for the undeveloped and the under-developed oil provinces of the third world to be combed for their resources over the next decade and would permit the development of large-scale productive capacity to be financed in cases where the reserves discovered, as well as other factors in the national situation, justified such development.

One possible strategy for this programme involves an intergovernmental agency, embracing both OECD and OPEC members, together with the utilization of the expertise of the international oil companies to ensure the technical success of the schemes. The world banking industry, given its close connections with the international oil industry, might be able to evolve alternative mechanisms whereby the aims and objectives of the new institutional arrangements could be achieved. Various options for creating the right combination of resources need to be examined in order to find a viable and acceptable response to the opportunity offered to the future of the international economic system by the developable, but so far largely undeveloped oil and gas resources of the developing world.

New oil reserves and production potential from the developing countries are needed to complement those which already exist in the OPEC countries as a result of the historical activities of the oil industry and those which are now being found in other, geographically limited, parts of the world, such as the North Sea and Australia, where the existing international oil companies are able to operate effectively. What is at stake, in responding to the challenge of the geographically more dispersed pattern of oil and gas exploitation on a large scale, is the establishment of conditions for economic progress and for enhanced and more widespread prosperity in the world system.

3

PETROLEUM EXPLORATION TRENDS IN THE DEVELOPING COUNTRIES

Jean Favre and F. M. Le Leuch
Institut Français du Pétrole

SPEEDING UP PETROLEUM EXPLORATION

The need to increase petroleum exploration efforts in the world so as to discover new reserves no longer has to be demonstrated on account of the tensions that have appeared in recent years on the international oil market, and which should become greater in the decades to come. These tensions, even if partly explained by political motivation, are due to an increasing gap between the supply and demand for oil and natural gas.

Indeed, all world energy forecasts agree on three conclusions:

(a) Between now and the end of the century, energy supplies will tend to be insufficient to sustain reasonable economic growth.
(b) The massive development of nonpetroleum energy sources will be much slower than initially foreseen.
(c) Recourse to oil will remain indispensable, and this will continue to be the main form of energy even if its relative share decreases (about 30% of energy supplies in the year 2000 instead of 47% in 1979, plus 18% for natural gas). To prevent too great tensions in the energy market, world oil production in the coming 10 to 20 years would have to increase by at least 30%, and that of natural gas would have to double.

Will it be possible to attain such production levels? Many recent surveys come to negative conclusions. Large hydrocarbon resources exist, but in many cases their discovery and exploitation come up against political, technical and economic constraints. Only a considerable speeding up of exploration investment efforts, on a worldwide scale in the next few years, would make it possible to attain a discovery rate higher than the production rate, as opposed to what has been observed since 1970 for the first time in petroleum history.

For the developing countries, the petroleum situation is even more disturbing since, outside of some 25 exporting countries, about 100 other countries must each year import increasing amounts of oil or petroleum products in order to maintain economic development. The sextupling of the price of oil in constant money has brought about a serious deterioration in their balance of payments since 1974. The only way for these countries to reduce their financial deficits, without slowing down their economic growth too much, is to develop their own potential hydrocarbon resources, which leads to a clearcut intensification of exploration activities in the oil-importing developing countries, where the growth of such activities should be greater than in the other parts of the world.

These observations illustrate the great importance of the Meeting on Petroleum Exploration Strategies in Developing Countries, and we thank the United Nations for having sponsored such a meeting of experts. The conclusions we shall reach at the end of our discussions will, we hope, help to speed up exploration efforts in the developing countries. This report will attempt to answer the following main questions:

(a) What is the current petroleum situation in the developing countries in terms of hydrocarbon supply and demand?
(b) How have exploration activities evolved in the developing countries in the last decade?
(c) What have been the exploration results in the developing countries and have any new exploration strategies been observed in recent years?

THE CURRENT PETROLEUM SITUATION OF THE DEVELOPING COUNTRIES: SUPPLY AND DEMAND

In order to assess exploration needs more accurately, it is essential to make a brief review of the petroleum situation in the developing countries by classifying them in groups of homogeneous countries as a function of their petroleum dependency (see Table 1):

(1) The oil-exporting developing countries, including both the current members of OPEC (13 countries) and other exporting countries (14 countries in 1980).
(2) The oil-importing developing countries, totalling about 100, including some which already produce oil (13 countries) or natural gas.

Naturally, any classification of this type cannot be perfect and must be updated because the petroleum situation of a country is linked to the evolution of supply and demand. Nonetheless, experience shows,

Table 1
Petroleum situation in the developing countries in 1980 (excluding centrally planned economies)

Oil-exporting developing countries				
Region	OPEC (13 countries)		Non-OPEC (14 countries)[a]	
Latin America	Ecuador Venezuela		Bolivia Mexico	Peru Trinidad and Tobago
Africa	Algeria Gabon	Libya Nigeria	Angola Cameroon Congo	Egypt Tunisia
Asia	Indonesia Iran Iraq Kuwait	Qatar Saudi Arabia United Arab Em.	Bahrain Brunei Malaysia	Oman Syria

Oil-importing developing countries			
Region	Producing (13 countries)[b]		Non-producing (80 countries)
	0—50%[c]	50—100%	
Latin America	Brazil Chile Guatemala	Argentina Colombia	19 countries
Africa	Ivory Coast	Ghana Zaire	40 countries
Asia	India Pakistan Philippines Turkey	Burma	21 countries

a Excluding countries with a current small exporting capacity (Burma, Zaire).
b Countries having a slight oil production (Barbados, Cuba, Morocco), a gas production (Afghanistan, Bangladesh, Thailand) or potential discoveries are not taken into account.
c Share of national production in 1980 oil consumption.

even in the case of significant discoveries, that several years are required before an oil-importing country can become an exporter.

The group of oil-importing developing countries is by far the largest since it represents a population of 1700 million inhabitants, i.e. 40% of the world population (compared with 500 million inhabitants in the exporting developing countries), and they consume only 10% of world oil consumption at present, of which four-fifths must be imported. Such figures demonstrate the need to assess hydrocarbon resources that may exist in the oil-importing developing countries.

Production of oil and natural gas

If the OPEC countries are excluded, the share of the developing countries in world production in 1979 amounted to 8.1% for oil and 5.2% for natural gas, as shown in Table 2.

(a) From 1960 to 1979, oil production by non-OPEC developing countries rose from 57 million t/year to 262 million t/year, i.e. a mean growth rate of 8.4% per year, which is higher than that of worldwide production (6.1%). This explains the increase in the share of the non-OPEC developing countries, which has risen from 5.4% to 8.1% of the world total as a result of the encouraging discoveries that have been made. For natural gas, the growth in relative values was even greater. The increase in marketed gas production has been more than 12% per year since 1960, with the volume rising from barely 10 billion m^3 /year

Table 2
Evolution of hydrocarbon production in the non-OPEC developing countries (DCs)

Region	Oil production (millions of t)				Marketed production of natural gas (billions of m^3)
	1960	1970	1979	1980[a]	1979
Non-OPEC exporting DCs	33.5	92	200	231	48
Latin America	23.2	33.7	103	133	34
(Mexico)	(14)	(22)	(80)	(110)	(28)
Africa	3.4	26	46	48	1.5
Asia	6.9	32.3	51	50	12.5
Importing DCs	23.5	52.5	62	63	30.5
Latin America	21.8	40.7	41	45	14.5
Africa	0.1	0.1	1.5	1.5	0.1
Asia	1.6	11.7	19.5	16.5	15.9
Total DCs	57	144.5	262	294	78.5
World production	1051	2335	3215	3065	1500
Share of DCs (%)	5.4	6.2	8.1	9.6	5.2

a Preliminary estimates.

to 80 billion m^3/year, or the equivalent of 80 million metric tons of oil. In absolute value, this production level remains low (5.2% of world production) as the result of difficulties in marketing natural gas, although this impact is tending to decrease because of new prices for energy.

(b) The situation with regard to oil-importing developing countries is more modest since for these countries, including several in Latin America with old fields being exploited, the growth rate of oil production (5.5% per year) since 1960 has hardly been equal to the world average. With an oil production of 62 million tons in 1979, they represent 2% of world oil production. For natural gas being marketed (30.5 billion m^3 in 1979), the share is exactly the same. The most disadvantaged region is still Africa (see Table 2).

(c) Generally speaking, 80% of the increase in oil production in the non-OPEC developing countries during the last 20 years has come from the group of countries already exporting, while the relative position of the oil-importing developing countries has not improved despite the appearance of new producing countries.

Proved reserves

The share of the non-OPEC developing countries in proved reserves is slightly greater than their share of production. With proved reserves of 9.6 billion tons (3.6 billion tons outside of Mexico), these countries account for 10.8% of all proved reserves (see Table 3), including only 1.2% for importing developing countries.

Table 3
Estimate of proved reserves in the non-OPEC developing countries (DCs) (situation end-1980)

Region	Oil (millions of t)		Natural gas (billions of m^3)	
Non-OPEC exporting DCs	8500	(42.5)[a]	4000	(83)
Latin America		6200		2500
(Mexico)		(6000)		(2000)
Africa		1000		500
Asia		1300		1000
Importing DCs	1100	(17.5)	2100	(68)
Latin America		670		895
Africa		30		5
Asia		400		1200
Total non-OPEC DCs	9600	(36.5)	6100	(77)
World Reserves	89000	(28)	75000	(50)
Share of DCs (in %)		10.8		8.1

a Proved reserves/1979 marketed production ratio expressed in number of years.

(a) Most of these reserves are actually situated in countries that are already exporters, i.e. 8.5 billion tons (including about 6 billion tons in Mexico) which, at the present rate of production, corresponds to 42 years of supply. If Mexico is excluded, this ratio drops to 21 years, which is less than the world average of 28 years, showing that, for this group of exporting countries, the growth of oil production will be possible only if extensive exploration efforts are undertaken to discover new commercial reserves.

(b) As to importing countries, they have proved reserves of 1.1 billion tons, i.e. 17.5 years' supply of production. It should be noted for these countries that the energy content of their proved reserves of natural gas (2100 billion m^3) is greater than that of their proved oil reserves. Furthermore, these gas reserves, most of which have only recently been discovered, have hardly been exploited because they correspond to a 68-year supply of marketed production at the present rate of exploitation, compared with a world average of about 50 years.

(c) Naturally, there is no way of using present proved reserves to make an accurate evaluation of the true hydrocarbon potential of the non-OPEC developing countries. They are the result of the exploration efforts that have been made to date. The too-low intensity of such efforts, and even their quasi-absence in a great many basins, show that the ultimate reserves of the non-OPEC developing countries are considerably greater than the present proved reserves. Whereas for the world the average of ultimate reserves is estimated to be at least 3 times the present proved reserves, the factor is considerably higher in the group of countries analysed.

Growth of oil imports

As things stand now, the non-OPEC developing countries as a whole have an oil deficit of about 150 million tons since their production 262 million tons in 1979) covers barely 65% of their consumption of petroleum products (410 million tons).

(a) Nonetheless, the situation is quite different between:

(i) Exporting countries, which have a surplus of 100 million tons, i.e. 50% of their production (production of 200 million tons in 1979 compared with a consumption of 100 million tons), providing them with revenues of $25 billion;

(ii) Importing countries, whose oil deficit is already 250 million tons, i.e. the equivalent of $65 billion at present oil prices.

(b) For these oil-importing developing countries, the volume of oil imports has increased considerably since 1960, as shown in Fig. 1. In 1979, imports amounted to 80% of oil consumption (310 million tons), with the remainder being supplied by domestic production (62 million tons). This oil deficit is already large, whereas the per capita consumption of petroleum products (0.18 TOE/year) for these countries is only one-fifteenth of the consumption in the developed countries.

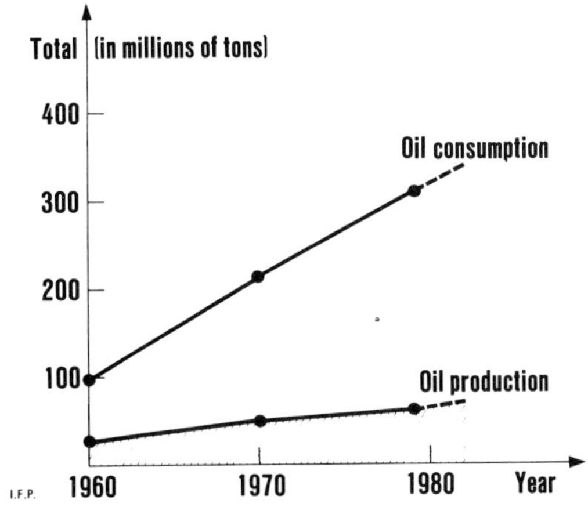

Fig. 1. Evolution of the petroleum deficit of the oil-importing developing countries (DCs)

(c) The effects of the oil deficit on the balance of payments of oil-importing developing countries have been considerable in recent years, and this trend will unfortunately continue in the years to come. Indeed, between now and the end of the century, the oil demand by the oil-importing developing countries could almost double, meaning that their average rate of growth would be 3.5% per year. Such a scenario would correspond to a slowing down of economic development and would imply a policy aimed at energy conservation (for which the possibilities, in absolute terms, are limited in the developing countries) combined with the increasing use of energy resources other than oil. Under such conditions, it is easier to understand the challenge raised by the intensification of oil exploration in the oil-importing developing countries. New hydrocarbon resources must be discovered so as to put the brake on the probable growth of oil imports beyond their present level of 250 million t/year, without bringing about any halt to the increase in petroleum needs for the sake of economic development.

EXPLORATION EFFORTS IN THE DEVELOPING COUNTRIES

Before analysing the results of exploration operations already undertaken, two important questions arise: has there been any speeding up of prospection efforts in the non-OPEC developing countries since 1973, and what proportion of world exploration is being done in these countries?

How to measure exploration efforts

From a theoretical standpoint, the most representative measurement of

Table 4
Estimate of exploration efforts in the world

Zone	Party-months of seismic prospecting				Exploratory wells			
	1970–74		1975–79		1970–74		1975–79	
	Number	%	Number	%	Number	%	Number	%
Developed countries	21000	64.5	26500	70.3	48800	89.7	64650	92.4
North America	17000	52.2	22500	59.7	47100	86.6	62650	89.5
Other countries	4000	12.3	4000	10.6	1700	3.1	2000	2.9
Developing countries	11600	35.5[b]	11200	29.7[b]	5574	10.3[b]	5338	7.6[b]
OPEC	4600	14.1	4650	12.3	2431	4.5	2377	3.4
Non-OPEC exporters	3050	9.3	2925	7.8	1374	2.5	1227	1.8
Already producers	2800	8.6	2825	7.5	1510	2.8	1492	2.1
Other countries	1150	3.5	800	2.1	259	0.5	242	0.3
Total[a]	32600	100	37700	100	54374	100	69988	100

a Outside of centrally planned economies, and excluding a few countries for which data are not available (Afghanistan, Burma, India, Syria).

b On account of the intensity of exploration in the developed countries, the cost of operations is lower there. The breakdown of exploration efforts measured in terms of expenditures is thus slightly less unfavourable for the developing countries.

exploration efforts would be to make a country-by-country estimate of the annual costs of such operations. Unfortunately, apart from the figures published by a few countries, there are as yet no homogeneous statistics indicating the financial effort devoted to petroleum exploration in each country. It should be noted that, generally speaking, the exploration sector is one of the aspects of the petroleum industry about which the fewest statistics exist. (Outside of a few specialized magazines, one of the best sources of information today is the STATSID data base, created by CFP, IFP and SNEA.) Listing the party-months of seismic prospecting or the features of the wells spudded each year in some 100 countries is actually more difficult than reporting the production figures. At the same time, several countries, for political reasons, continue to consider such information confidential. Progress will doubtless be made in the future with regard to exploration statistics.

To illustrate the evolution of exploration operations since 1970, we shall give estimates of both the number of party-months of seismic prospecting and the number of exploratory wells drilled. Interpreting the results is relatively difficult on account of the diversity of the operations undertaken. As an example of this, the depth of an exploratory well may vary from 1000 to 6000 metres, and the cost of a well, depending on the geographic conditions, may fluctuate in a ratio of 1 to 50. Similarly, this approach cannot be used for directly measuring the movement of exploration towards more and more difficult zones, requiring enormous budgets to be earmarked for each such operation.

Stability of exploration efforts in the developing countries

Table 4 shows that, in the developing countries, exploration efforts, including both seismic prospecting and drilling, have remained almost stable if the work done during each five-year period (1970–74 and 1975–79) is compared.

(a) In the last ten years, an average of 2250 party-months of seismic prospecting have been carried out each year in the developing countries, broken down as follows: 41% in the OPEC member countries, 26% in other exporting countries and 33% in oil-importing developing countries, including 25% (nearly 600 party-months/year) in countries already producing and 8% (200 party-months/year) in countries not yet producing. Drilling tended to decrease in the developing countries but this represents only a small proportion of seismic prospecting (see Fig. 2).

(b) The same conclusions can be drawn for exploratory wells. On an average, 1100 wells have been drilled per year in the developing countries since 1970, with 43% in the OPEC member countries, 24% in the other exporting countries and 33% in the importing countries, including 28% (300 wells/year) in countries already producing, and barely 5% (50 wells/year) in the other 80 importing countries, i.e. an average of about 0.6 wells/year/country for these countries as a whole.

(c) It can be seen that seismic prospecting efforts are slightly more pronounced, in relative value, than drilling efforts in the importing

Table 5
Evolution of the number of exploratory wells in the developing countries (DCs)[a]

Zone	1970	1971	1972	1973	1974	1975	1976	1977	1978	1979	Total 1970–79	
OPEC countries	386	510	523	501	511	469	454	434	491	529	4808	(43%)
Non-OPEC exporting DCs	286	285	308	243	252	245	205	232	273	272	2601	(24%)
Oil-importing DCs	358	369	326	357	359	346	350	386	346	306	3503	(33%)
Latin America	271	289	250	264	267	225	247	300	231	227	2571	(24%)
Africa	46	32	28	25	29	35	26	30	27	26	304	(3%)
Asia-Oceania	41	48	48	68	63	86	77	56	88	53	628	(6%)
Total DCs	1030	1164	1157	1101	1122	1060	1009	1052	1110	1107	10912	(100%)

a Excluding a few countries for which data are not available.

countries. This can be explained by the fact that little information exists on the petroleum potential of these countries and that seismic prospecting, after geology, is the best method for gaining more knowledge of new basins and for reducing risks with regard to subsequent wildcats. Nonetheless, only by the drilling of wildcat wells is it possible to evaluate the actual hydrocarbon potential of a zone, and it is easy to see the need of encouraging such operations despite the risks associated with them.

(d) More detailed statistics for exploration drilling undertaken since 1970 in the developing countries are summarized in Table 5. They confirm the relative stability of the number of wells during the period and the small number of operations undertaken in the 100 oil-importing developing countries, especially in Africa (some 30 wells per year) and Asia-Oceania (some 60 wells per year).

(e) Therefore, contrary to what might have been expected, the 1973 oil crisis did not bring on, six years later, a speeding up of exploration efforts in the developing countries if these efforts are measured in terms of party-months of seismic prospecting or the number of wildcats

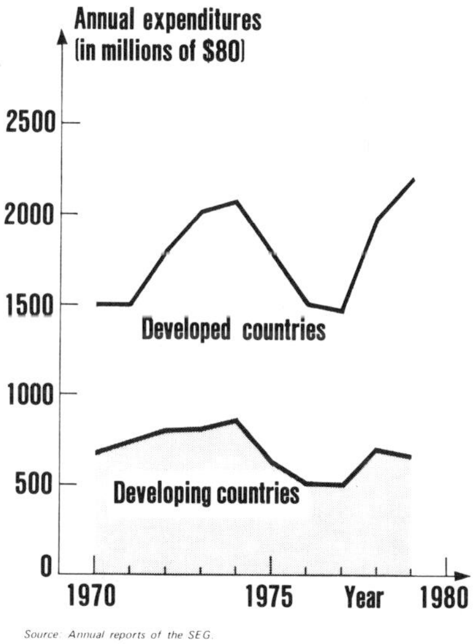

Fig. 2. Evolution of seismic prospecting expenditures (in 1980 dollars).

drilled. Naturally, with regard to exploration budgets, expenditures have appreciably increased on account of the increase in individual exploration costs, which has been greater than inflation, especially for drilling, and the carrying out of operations in more and more difficult zones (isolated zones, offshore basins).

Table 6
Evolution of the number of exploratory wells in the world[a]

Zone	1970	1971	1972	1973	1974	1975	1976	1977	1978	1979	Total 1970—79	
North America	9250	8475	9300	9700	10375	10880	11700	12770	13800	13500	109750	(88%)
Other developed countries	349	307	358	340	346	400	355	340	455	450	3700	(3%)
OPEC countries	386	510	523	501	511	469	454	434	491	529	4808	(4%)
Other DCs[b]	644	654	634	600	611	591	555	618	619	578	6104	(5%)
Total	10629	9946	10815	11141	11843	12340	13064	14162	15365	15057	124362	(100%)

a Outside of centrally planned economies.
b Excluding a few countries for which data are not available.

(f) It is still too early to see from the statistics whether the new increases in the price of oil in 1979-80 — increases which were greater in absolute value than the ones in 1973-74 — have had a stronger impact on exploration budgets in the developing countries. Nevertheless, several factors, such as the determination expressed by oil companies to take out new exploration permits in this group of countries and the beneficial competition resulting from this, suggest that it will really be in the next few years that the true intensification of petroleum prospection will finally take place in the developing countries. The exploration aid programme decided upon in January 1979 by the World Bank has certainly been a favourable factor in this evolution.

Comparison with exploration trends in the developed countries

Three main conclusions can be drawn from the exploration efforts made in the world (see Table 4):

(a) Most exploration operations have been undertaken in the developed countries, since the developing countries represent only 30% of the seismic prospecting and barely 10% of the wildcats drilled (excluding the centrally planned economies).
(b) Since 1975, an appreciable speeding up of exploration efforts has taken place in the developed countries, contrasting with the stability seen in the developing countries (see Table 6).
(c) Because of specific political and economic circumstances, exploration operations, especially with regard to drilling, have mainly been concentrated in North America (United States and Canada). On the other hand, more prospection efforts have been made in the developing countries than in the developed countries, not counting North America. Therefore, the 15 000 exploratory wells drilled in the last 10 years outside of North America break down into 25% in the developed countries and 75% in the developing countries, of which 40% were in the non-OPEC developing countries.

If the intensity of exploration is compared with the sedimentary zones liable to contain oil (excluding deep offshore areas), it can be seen that the non-OPEC developing countries with nearly 25 million km^2, representing an area slightly larger than that of the developed countries, including North America, are highly disadvantaged.

Why is the growth of exploration limited to the developed countries?

The speeding up of petroleum prospection in the developed countries can largely be explained by economic and political factors, since the geological risk linked to the very nature of the operations also exists there and, to a certain extent, is sometimes even greater.

(a) Considering the number of wells already drilled in several developed countries (e.g. North America), a decrease in exploration yields can be seen. The size of discoveries is decreasing, and there is less chance of making a discovery than there used to be. The continuing decrease in the amounts discovered per metre drilled in the United States or even in the North Sea is a perfect illustration of this phenomenon, which is characteristic of the basins already prospected on a large scale.

(b) The intensification of exploration in the developed countries is thus mainly due to new economic conditions resulting from extensive rises in the price of oil which may justify costly exploration projects if the tax structure is suited for such cost levels. An economic analysis effectively shows that current oil prices are such that, in the tax structures in force in various developed countries, oil companies are in a position to undertake costly exploration projects capable of costing $15 (1980 dollars) per barrel, which is higher than the price of oil just two years ago. The example of the United States is typical of this.

(c) The existence of fewer political risks may also justify the priority earmarking of exploration expenditures in the developed countries. Nevertheless, repeated adjustments of the tax structure in these countries (windfall profits tax in the United States, petroleum revenue tax in the United Kingdom, special tax in Norway, etc.) prove that conjunctural adaptations of the tax structure are not the prerogative of the developing countries. On the contrary, several of the latter offer a high degree of stability in the conditions proposed.

At the same time, it can be seen that current exploration levels in North America are remaining lower than the maxima seen 25 years ago (see Fig. 3), thus proving that exploration expenditures could continue to rise in these countries, even if the yield of such exploration is relatively low.

How can exploration in the developing countries be encouraged in the future so as to increase their share in worldwide exploration? Whereas it is almost impossible, as in the developed countries, to bring

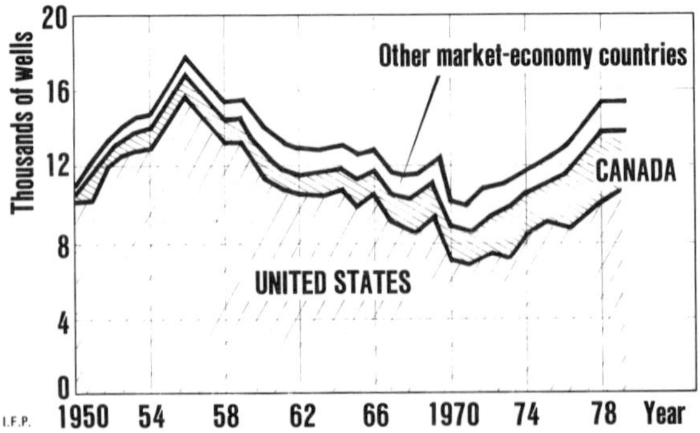

Fig. 3. Evolution of the number of exploratory wells drilled in the market-economy countries.

about any technical reduction in geological exploration risks, the developing countries can adopt new incentives of an economic or political nature to attract a larger proportion of exploration budgets.

On the basis of information published by some countries, together with estimates based on seismic prospecting or drilling costs, Table 7 gives an evaluation of the exploration expenditures made in 1979 in the market-economy countries.

(a) It can be seen that, out of the $14 billion invested in petroleum prospection in 1979, $3.4 billion was earmarked for the developing countries, i.e. one-quarter of total expenditures. Of this total, about $1.1 billion went to the 100 or so oil-importing developing countries, including only $300 million, or 2% of the worldwide exploration budgets, to the 80 or so countries in which no commercial production exists as yet.

(b) Whereas in the developed countries the share of the international oil companies is preponderant, this share in the developing countries,

Table 7
Estimate of exploration budgets in 1979[a] (in millions of dollars)

Region	Expenditures for geology, geophysical prospecting and miscellaneous	Expenditures for drilling	Total exploration expenditures	Breakdown (%)
North America	2000	7000	9000	64
Other developed countries	350	1250	1600	11
Developing countries	850	2550	3400	25
OPEC countries	350	1150	1500	11
Non-OPEC exporting countries	200	600	800	6
Importing DCs	300	800	1100	8
Total[b]	3200	10800	14000	100

a Excluding rentals and bonuses paid to governments for the granting of permits (about $4 billion). Exploration expenditure figures indicate approximate order of magnitude only.
b Outside of centrally planned economies.

Table 8
Estimate of hydrocarbon discoveries in non-OPEC developing countries (DCs)

Commercial discoveries[a]	Oil				Natural gas		
	Exporting DC		Importing DC		Exporting DC	Importing DC	
> 1000	Mexico	(1900)[b]			Mexico		
500 — 1000	Egypt	(1938)	Argentina	(1907)	Malaysia	Argentina	
	Oman	(1962)	India	(1889)			
250 — 500	Angola	(1962)	Brazil	(1941)	Bahrain	India	
	Brunei	(1929)	Colombia	(1918)	Brunei	Pakistan	
	Malaysia	(1911)			Trinidad		
	Syria	(1956)					
	Trinidad	(1902)					
	Tunisia	(1964)					
100 — 250	Bahrain	(1932)	Chile	(1918)	Bolivia	Bangladesh	(1959)
	Peru	(1869)			Cameroon	Chile	
					Egypt	Colombia	
					Tunisia	Thailand	(1973)
50 — 100	Bolivia	(1927)	Turkey	(1940)	Congo	Afghanistan	(1971)
	Cameroon	(1972)			Oman	Brazil	
	Congo	(1969)			Syria		
< 50[c]			Burma	(1902)	Angola	Turkey	
			Ghana	(1977)	Peru		
			Guatemala	(1972)			
			Ivory Coast	(1976)			
			Morocco	(1934)			
			Pakistan	(1937)			
			Philippines	(1976)			
			Zaire	(1970)			

a In millions of t for oil; in billions of m^3 for natural gas. Commercial discoveries include both current proved reserves and cumulative production in the country.
b Year of first discovery.
c Excluding Barbados, Benin, Chad, Tanzania and Sudan.

considering the actual budgets earmarked directly by some national oil companies (Mexico, Brazil, India, Argentina, Turkey, etc.), drops on average to half of the expenditures. Nonetheless, outside of a few countries in which national companies are active, the international oil companies generally assume responsibility for the entire exploration risk. In 1979 these companies spent between \$1.5 and \$2 billion in all the developing countries, i.e. 15 to 20% of their worldwide exploration budgets. The European companies made a comparatively greater effort with regard to the developing countries.

MAIN RESULTS AND EXPLORATION STRATEGY

The exploration operations undertaken in the developing countries have mainly concentrated on zones that are already productive, even if, in recent years, there has been a greater geographic dispersion of exploration activities. This policy has led to the discovery of new hydrocarbon-rich basins, thus demonstrating the importance of undertaking prospection in zones which are not yet very well known from the petroleum standpoint.

Hydrocarbon discoveries in non-OPEC developing countries

As of now, more than 13 billion tons of oil (and 7500 billion m^3 of gas) have been discovered in the non-OPEC developing countries, i.e. 9% of present worldwide discoveries (150 billion tons). More than 80% of these reserves were discovered after 1960 in the developing countries, which is a much higher proportion than in the rest of the world. Table 8 gives the relative size of discoveries by country.

(a) As expected, most of these discoveries were made in developing countries which already export oil (including Mexico). Besides countries which have had oil for a long time (Bahrain, Bolivia, Brunei, Egypt, Malaysia, Mexico, Peru, Trinidad and Tobago), where new basins are continuing to be found, five countries (Tunisia in 1966, Oman and Syria in 1968, the Congo in 1969, and Cameroon in 1978) with discoveries of nearly 1.5 billion tons have become producers and exporters. On the other hand, some countries such as Bolivia and Peru could once again become oil importers if new fields are not found.

(b) A total of 2.3 billion tons of oil (1.6% of worldwide discoveries), which is slightly less than the present proved reserves in the North Sea, has been discovered in the oil-importing developing countries as a whole. Since 1960, the average rate of discoveries has been about 70 million t/year, which is slightly higher than production in 1979. Most of these discoveries were made in eight old producers: Argentina, Brazil, Burma, Chile, Colombia, India, Pakistan and Turkey. Five new producing countries have also appeared in recent years: Zaire in 1975, Guatemala in 1978, the Philippines and Ghana in 1979, and the Ivory Coast in 1980. Discoveries in these countries to date are still limited to about 100 million tons, but they might be re-evaluated, especially in the Ivory Coast. Potential oil discoveries have also been made in other countries (Benin, Chad, Sudan, etc.).

(c) It should be noted that discoveries of natural gas in the oil-importing developing countries amount to 2500 billion m^3 (2.5% of worldwide discoveries), most of which are situated in Argentina, Bangladesh, Chile, Colombia, India, Pakistan and Thailand.

Growing production of offshore oil

Since 1960, offshore petroleum activities have undergone rapid development in the world since, as things stand now, offshore production accounts for nearly 25% of the petroleum production by the market-economy countries.

(a) This trend has taken on even greater intensity in the non-OPEC oil-exporting developing countries for which offshore production amounts to nearly 40% of their production on the average, particularly as the result of the resources exploited by Mexico, Trinidad and Tobago in Latin America, by Egypt and Angola in Africa, and by Brunei and Malaysia in Asia.

(b) On the other hand, this trend has moved at a slower place in the oil-importing developing countries, where offshore production (10 million t/year) still represents only 15% of their oil production, which comes mainly from fields lying offshore Brazil (since 1973), Zaire (1975), India (1976), Ghana and the Philippines (1979), Chile and the Ivory Coast (1980), without counting Benin and Thailand (natural gas). Nonetheless, the increase in oil production since 1970 in this group of countries stems solely from the contribution of new offshore fields, and this trend will increase even further in the light of present proved reserves. This can be explained by the fact that offshore sedimentary basins, lying underneath the continental shelf and the continental slope, have not been sufficiently explored to date despite the fact that they contain large-scale hydrocarbon resources. A great many countries have effectively opened up these zones for exploration only in the past few years.

Geographic concentration of exploration efforts

The location of oil prospection operations shows that most developing countries have effectively had, at a given time or even continuously, operations undertaken on their national territory. A more thorough analysis of the period from 1970 to 1979 shows that such exploration is actually highly concentrated in countries that are already producers, as shown in Fig. 4. One-fourth of the developing countries account for 90% of the exploratory wells drilled.

Consequently, any programme to intensify exploration must try to encourage operations in countries where no significant discovery has been made as yet, not because of the absence of any petroleum potential, but solely because prospection has been too scattered.

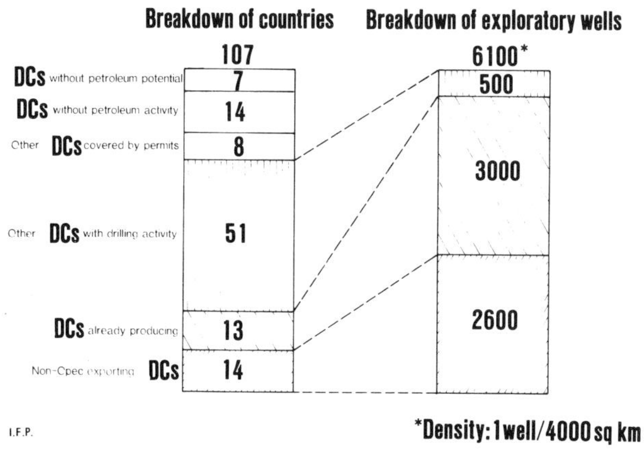

Fig. 4. Concentration of exploration efforts in non-OPEC developing countries (DCs) (1970–1979).

(a) If we exclude the several countries where there is almost no probability of finding oil (7 African countries), we find that there are some 70 countries having a petroleum potential and in which exploration efforts, despite the drilling of wildcats or seismic prospecting, have been too sparse to gather all the basic information required for evaluating their veritable hydrocarbon potential. As things stand now, no exploration permit is in force in about 20 of these countries.

(b) Concerning the geographic diversification of exploration, two trends must continue to receive favoured treatment.

(i) Increased prospection on the continental shelf where technical costs, outside of seismic prospecting, rise very quickly with water depth, thus requiring the terms of contracts to be formulated so as not to discourage exploration.

(ii) The assessment of numerous onshore basins which have remained virtually unexplored on account of technical or economic constraints (landlocked countries, isolated zones, deep-lying basins, etc). As an example, the start of prospection has been observed in recent years in basins in the heart of Africa (Sudan, Chad, Niger, etc.), and encouraging results have already been obtained.

This orientation must not make us forget that, in the oil-importing developing countries that are already producers, where domestic production covers only a portion of the petroleum needs, the speeding up of prospection should also be encouraged. It must not be forgotten that in only about 10 oil-importing developing countries more than 100 wildcat wells have been drilled (Morocco in Africa; Argentina, Brazil, Chile and Colombia in Latin America; and India, Pakistan and the Philippines in Asia).

It can be pointed out that several countries having vast sedimentary basins (Argentina, Brazil, India, etc.) have already decided to promote this policy of encouragement by making a more widespread appeal for co-operation by the international oil companies. Many of these companies have also decided to work on a larger scale in the developing countries, as demonstrated by their presence when new permits are being given out. Another example outside the oil-importing developing countries confirms the fact that oil companies are attracted by the idea of prospecting in relatively virgin sedimentary basins. Such is the case of China.

Are sufficient exploration efforts being made in the developing countries?

There are no absolute rules for determining the optimum exploration efforts which should be undertaken in the developing countries. It depends on the hydrocarbon potential of each country, which is difficult to estimate, on the proportion of the petroleum needs of the country and on the eventual desire of the country to become an exporter.

(a) For the oil-importing developing countries as a whole, petroleum independence would mean that current production (62 million t/year) would have to be multiplied fivefold, not including the foreseeable increase in consumption. A less ambitious objective for increasing production, and also a more realistic one, by itself justifies the speeding up of prospection, especially since present stocks of proved reserves (17.5 years) do not encourage a sharp rise in production, except for natural gas.

(b) What are the limits to be recommended for an exploration programme? Experience in various developed countries shows that discovery costs of $15/barrel (in 1980 dollars) may be acceptable if the tax conditions are favourable, whereas until recently discovery costs were a few dollars per barrel. This new change of scale means that exploration strategies must be looked at from a different angle. Petroleum objectives that were abandoned several years ago are now becoming economically feasible (small offshore fields, inland basins, etc.). For all the non-OPEC developing countries, where average discovery costs are less than $3/barrel, it is easy to see that exploration should be intensified.

The effort to discover hydrocarbon resources requires tenacity. The drilling of several dry wells in a given basin does not lead to any certain conclusion as to the absence of any petroleum potential.

(a) Examples of failures by initial exploration campaigns in a region, followed by commercial discoveries, are numerous: Ecuador, Libya, the North Sea, Cameroon, the Philippines, etc. Before oil is discovered in a zone, substantial exploration budgets must often be earmarked, sometimes amounting to several hundreds of millions of dollars, and this is true despite the improvements made by petroleum technology.

(b) Such exploration risks must be taken into consideration in the contractual aspects if foreign interests are to be encouraged to participate, and these interests must be assured of being justly remunerated for the risk capital they invest.

The leading factor which may slow down the expansion of prospection activities in the oil-importing developing countries in the coming years could very well be the difficulty in financing high-risk capital by oil companies, both national and international. Outside of geological risks, these investments will be linked directly to the conditions that are offered.

(a) It may be expected that in countries with oil production such as Argentina, Brazil and India, where national oil companies have increased their own exploration budget, transnational oil corporations could increase their exploration expenditures in the future.

(b) For many countries where there is as yet no production, the international oil companies will probably have to account for most of the exploration investments, outside of bilateral or multinational financing possibilities.

RECOMMENDATIONS: EXPLORATION INCENTIVES

The extent of exploration in the developing countries will depend, to a large degree, on the petroleum policy that is adopted by each sovereign state. A great many examples clearly demonstrate the impact of petroleum policy on exploration activities.

Those developing countries desiring almost all of the work to be done by their national company will have to make sure that this company obtains the financing required for doing more and more expensive jobs and for making use of the most effective technology. Some national companies in oil-importing developing countries already have annual oil financing needs (including exploration and development expenditures) which are higher than one billion dollars. Considering the importance of speeding up exploration in the developing countries and the scope of the tasks to be accomplished in prospective areas that are often larger than several hundreds of thousands of km^2 per country, it can readily be seen that co-operation with international oil companies appears to be an appropriate solution. Moreover, several recent decisions demonstrate that this policy has often been pursued with success. To facilitate such co-operation, a favourable climate for the long-term commitment of foreign capital will have to be fostered.

Depending on the degree of petroleum development of the country, the procedures for attracting foreign investments may be different:

(a) In countries where discoveries have already been made and where exploration is being pursued, the immediate granting of permits accompanied by equitable conditions for both parties is the normal solution.

(b) On the other hand, in countries where exploration has been limited, which are by far the most numerous, the intensification of exploration also leads to the need for a promotion campaign so as to determine the potential significance of the new zone to be explored. This strategy implies the carrying out of the following main phases (see Fig. 5):

(i) Evaluating, with the help of specialized consultants, the hydrocarbon potential of the country on the basis of existing information or the results of additional exploration (geology, seismic prospecting or even stratigraphic drilling, etc.).

(ii) Preparing a legal and tax climate that is suited to the techno-economic conditions liable to be encountered in the zone to be prospected:
   (a) choice of type of contract;
   (b) setting the terms of contracts so that the conditions offered are attractive both for important discoveries and for small discoveries;
   (c) safeguarding the long-range interests of the country.

(iii) Awarding permits to oil companies wishing to undertake meaningful work, including wildcat drilling.

(iv) Participating in the work on the technical level and, if the country can obtain the necessary financing, on the financial level as well.

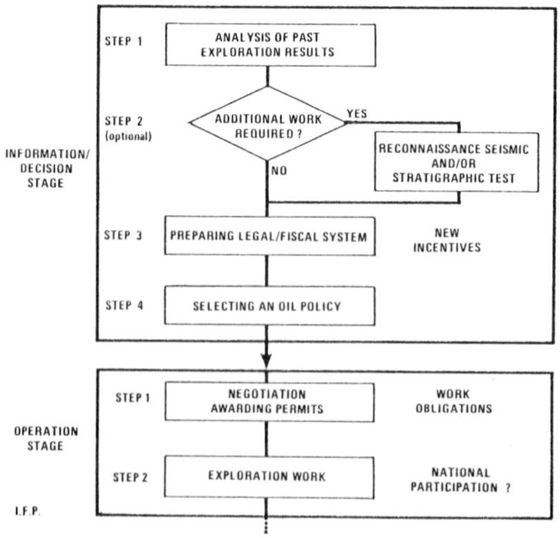

Fig. 5. Strategy to promote oil exploration (LDCs) with current limited exploration).

The exploration assistance plan adopted by the World Bank in January 1979 facilitates financing for promoting new petroleum zones by means of any loans (World Bank loans, IDA credits) that may be granted. Because this plan is very recent, the amount of the loans for exploration granted by the World Bank (outside of evaluation or development operations) is still low compared with the expenditures undertaken in this field. Their proportion will doubtless increase considerably. In the long run, a multinational financing system with conditional repayment could also be encouraged, because, unlike the present situation, future loans might be reimbursed only if a discovery is made. These different solutions are particularly aimed at financing the needs of developing countries or their national companies.

For the international oil companies, the above arrangements serve to limit the financial risks involved in exploration to a certain degree through the immediate financial participation of the government in the exploration operations (and not solely in the case of discovery, i.e. without assuming any prospection risks). Other tax measures may also be planned by the government, but they become effective only when production ensues. Furthermore, their impact depends on a guarantee of stable conditions during production. Nonetheless, such solutions must be investigated because the return on high-risk investments for exploration in the developing countries must be at least equal to what can be obtained in the developed countries. If this is not the case, the present breakdown of exploration budgets will remain.

CONCLUSIONS

In the light of the increasing burden of oil imports for the developing countries, for which present domestic production covers only 20% of the needs, as well as the resulting deficit in the balance of payments and its impact on the possibilities for economic growth, exploration assistance measures should be encouraged so as:

(a) to increase the petroleum resources available in countries that are already producers;
(b) especially, to ensure a better assessment of the hydrocarbon potential of some 70 other developing countries.

The ultimate reserves of the oil-importing developing countries could probably be at least five to ten times current proved reserves, and this in itself justifies such a policy. The discovery of these reserves will require considerably higher budgets than the ones now committed.

(a) The minimum medium-term objective should be to double these budgets by raising them from about $1.25 billion (1980 dollars) in 1979 to an annual average of $2.5 billion (1980 dollars) for the period 1982-85, which would represent only 4% of the current value of the petroleum imports by these countries.

(b) Such a goal can be attained if, at the same time, arrangements are planned to encourage oil companies to earmark a larger share of their investments for the developing countries and to help the governments to finance their share of the prospection operations.

4

EAST AFRICAN BASINS: RESERVES, RESOURCES AND PROSPECTS

Arthur Whiteman
Professor, Petroleum Exploration Studies, Aberdeen University, Scotland

INTRODUCTION

According to Coury and Henricks (1978) more than 60 sedimentary basins occur on the continental portion of the African lithospheric plate and more than 12 of these on one count (and at least 19 on another) occur along the eastern continental margin of Africa. These 60 sedimentary basins together occupy about 5×10^6 square miles and the Eastern African group of basins extending from the Cape of Good Hope to Cape Gardafui, Somalia (including onshore and offshore down to 200 m water depth) occupy just over two-thirds of a million square miles. A list of African basins and approximate areas are presented in Table 1. Their distribution is shown in Figs 1 and 2. Petroleum geology and prospects of East African basins are treated in summary form by country in Part II.

A comparison of basin shapes shown in Figs 1 and 2 demonstrates the highly subjective nature of basin definition and use. For instance, Coury and Henricks (1978) used a 1 km sediment thickness cut off and showed only producing basins and basins which they think have oil and gas potential; consequently they omit the Karroo basins of Zambia and adjacent areas but show the Karroo basin of South Africa which, even in official circles (South Africa's National Oil Company), is thought in large part to have no oil and gas potential. There are many other 'subjective' selections shown on this map. Grossling (1976) did not define cut offs except generally and selected 200 or 2000 m water depths for determination of limits to offshore basins. In some cases his estimates of prospective areas are too high. Eastern African basin areas in this paper are calculated as total onshore sedimentary areas plus those parts of the sedimentary prism extending down to 0 - 200 m and 200 - 3000 m water depth (Fig. 1).

Nevertheless, despite these imperfections and discrepancies, attempts

Table 1
List of basins situated mainly on the continental portion of the African lithospheric plate; 1 km thickness and 0.2 or 2 km water depth cut offs; and approximate surface areas

ONSHORE BASINS	ONSHORE-OFFSHORE BASINS	OFFSHORE BASINS	AREA SQ MI × 10^3
RHARB			4
ESSAOUIRA			34
		ALBORAN	13
		WEST MEDITERRANEAN	>10
ATLAS			77
	CHELIF		8
HODNA-CONSTANTINE			3
	CAPE BON		63
	TARFAYA-RIO DE ORO		67
TINDOUF + REGGANE			83
COLOMB BECHAR MACMAHON			94
GRAND ERG ORIENTAL			150
SIRTE			285
WESTERN DESERT			145
NILE DELTA			20
NILE BASIN			50
	GULF OF SUEZ TROUGH		11
	SINAI		25
	DEAD SEA PLATFORM		>9
	RED SEA		166
KUFRA			161
MURZUK			83
TANEZROUF			18
TAOUDENI			263
	SENEGAL		188
	BOVE		8
	SIERRA LEONE COASTAL		22
	LIBERIAN SHELF AND SLOPE		19
VOLTAIAN			39
IULLEMMEDEN			171
NORTH CHAD			120
CHARI			38
DOBA			13
SUDAN TROUGH			112
OGADEN			116
	SOMALIA		164
	LAMU		119
	IVORY COAST		55
	DAHOMEY COAST		16
BENUE TROUGH			45
	NIGER DELTA		48
	CAMEROUN COASTAL		18
	GABON COASTAL		54
CONGO			246
	CABINDA COASTAL		48
	CUANZA		36
	MOCAMEDES		10
ETOSHA			190
KALAHARI			84
	ORANGE RIVER		105
KARROO			84
		SOUTH AFRICAN COASTAL	29
	GAMTOOS		3
	ALGOA		7
	LIMPOPO		205
	MAJUNGA		26
	MORONDAVA		51
	TAMATAVE		29
	TANZANIA		88
		MADAGASCAR PLATFORM	13
		MAURITIUS PLATFORM	155
TOTAL MINIMUM NUMBER AFRICAN BASINS (>60)			4.614×10^6
TOTALS 27	29	4	

Source: Basins identified on 1:20 000 000 equal area map of Coury and Henricks (1978). Area estimates by A.J. Whiteman using simple squares method.

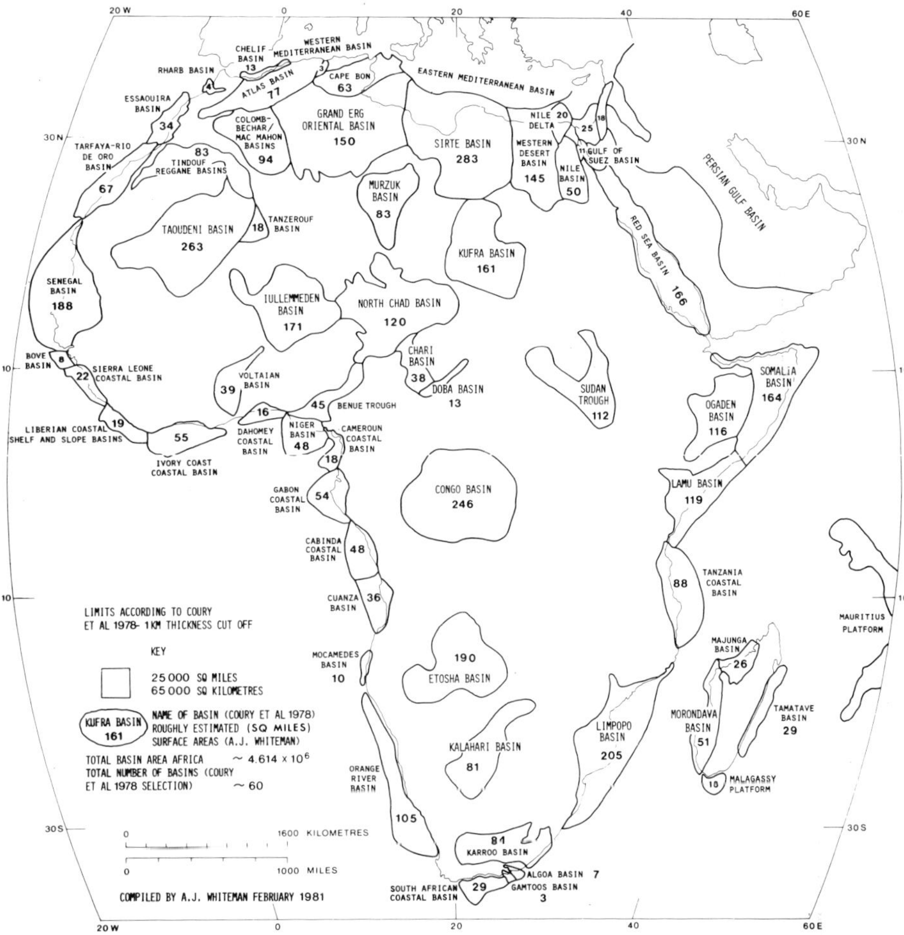

Fig. 1. Prospective basins: Africa.

to quantify and define African basins are valuable and assist us in building up ideas about prospective areas and reserve-production-resource inventories, in providing scale, and in drawing attention to the apparently abnormally low petroleum yields of Eastern Africa. It is unlikely that the quality of data available to us will change rapidly or dramatically, except perhaps for a few individual countries, and what information is now available publicly for most of Africa — much of it generated during colonial days — will form the greater part of our data base for many years to come.

Twelve out of the 60 African basins that have been recognized up to now are said to contain 'proven reserves' ('remaining recoverable reserves') totalling 55.1×10^9 bbl of oil and 208×10^{12} cu. ft of gas. These basins extend across at least 12 countries. The lion's share of the

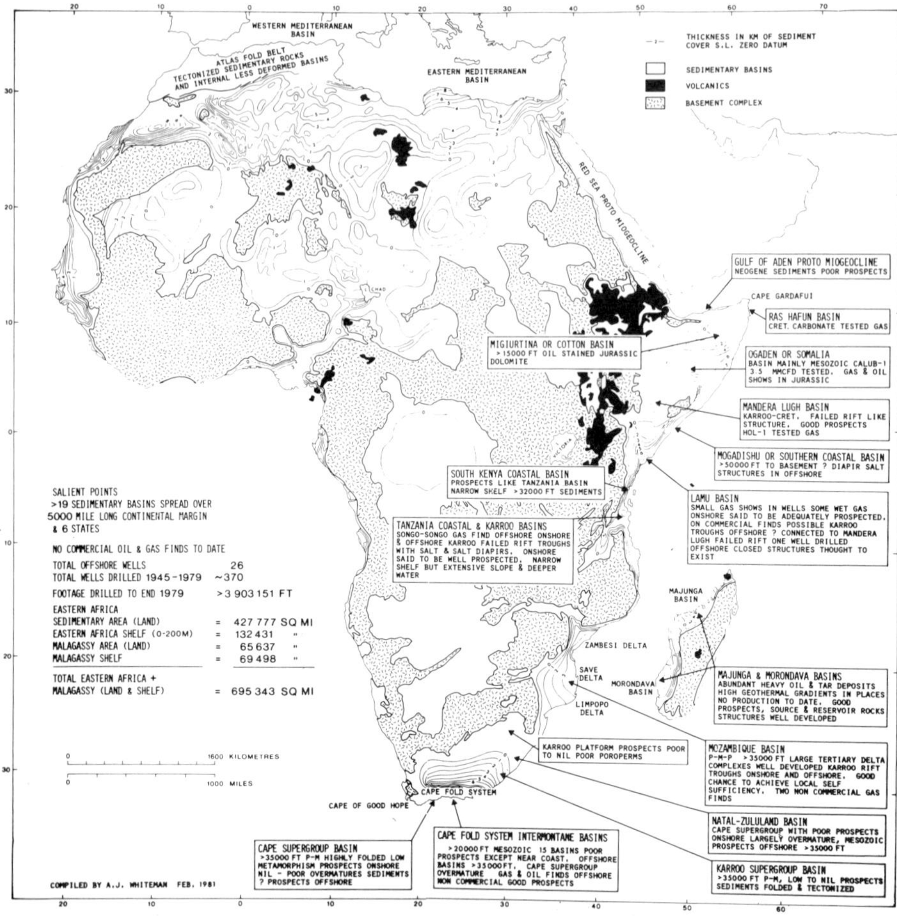

Fig. 2. Eastern Africa basins.

remaining recoverable oil and gas is located in Libya, Nigeria, Algeria and Egypt, while the states of Eastern Africa do not have any production as yet.

In the remaining 48 African basins (Fig. 2), oil and gas were discovered in some instances but produced in small quantities (Chad, Morocco) or production was sometimes shut down (Superior Oil's gas discovery in South Africa's Plattenberg Bay offshore basin, Gulf Oil's Pande gas discovery in the onshore Mozambique basin and the AGIP-AMOCO Songo Songo gas show offshore in Tanzania).

Grossling (1976) commented upon the proven skewed pattern of oil and gas distribution in Africa and gave reasons for it, pointing largely to the very low drilling density (0.0094 wells per square mile — Table 2) for Africa excluding Algeria, Egypt, Libya and Nigeria. Recent exploratory and development efforts have done little to change the general pattern to which he drew attention six years ago (Table 3).

Table 2
Prospective onshore and offshore areas for Africa and Madagascar (to end 1975)

COUNTRY	PROSPECTIVE AREA ONSHORE (SQUARE MILE) $\times 10^3$	OFFSHORE (SQUARE MILE) $\times 10^3$	TOTAL $\times 10^3$
ALGERIA	635	5.28	
LIBYA	543	32.2	
NIGER	390	-	
EGYPT	310	14.5	
MAURITANIA	290	17.0	
ZAIRE	270	0.4	
SUDAN	250	8.6	
MALI	250	-	
ETHIOPIA	220	18.0	
CHAD	220	-	
SOMALIA	172	15.0	187.0
NIGERIA	157	17.8	
MOZAMBIQUE	118	40.0	158.0
SOUTH AFRICA	95	55.0	150.0
MOROCCO	120	24.0	
MADAGASCAR	76	35.0	111.0
SPANISH SAHARA	61	23.0	
TUNISIA	63·4	19.5	
SENEGAL	53	12.0	
KENYA	45	5.5	50.5
ANGOLA	20	25.0	
TANZANIA	28	16.0	44.0
GABON	23	17.7	
SW AFRICA	30	5.0	
CONGO	27	3.5	
GUINEA-BISSAU	7	17.8	
GUINEA	9.5	14.8	
IVORY COAST	11	4.0	
SIERRA LEONE	4	10.2	
CAMEROON	6	4.1	
GHANA	2	8.05	
ZIMBABWE	9	-	
LIBERIA	0	7.5	
DAHOMEY	6·5	0.66	
GAMBIA	4	2.2	
LESOTHO	6	-	
SWAZILAND	5.4	-	
CENTRAL AFRICAN REPUBLIC	5	-	
EQ. GUINEA	1	4.0	
UPPER VOLTA	5	-	
TOGO	2	1.5	
SUB TOTALS	4 549.800	484.790	700.5
TOTAL (ON + OFFSHORE)	5 034 590 Sq. Miles		

Source: Grossling (1976).

PART I
EASTERN AFRICA BASINS: GENERAL

If we take a closer look at the Eastern African continental margin we can recognize at least 19 basins and a number of sub-basins (Whiteman, 1977). These include:

(1) Cape supergroup, Cape Province basin, South Africa
(2) Cape supergroup, Natal Province basin, South Africa
(3) Cape fold system intermontane basins, South Africa

Table 3
Estimated proven reserves, cumulative production, oil production (oil) and proven reserves (gas), producing countries, Africa

COUNTRY	ESTIMATED PROVEN RESERVES $\times 10^9$ BBL OIL	ESTIMATED PROVEN RESERVES $\times 10^{12}$ CUF GAS	CUMULATIVE PRODUCTION TO 1:7:80 $\times 10^9$ BBL OIL	EST. ORIGINAL RECOVERABLE OIL $\times 10^9$ BBL	PRODUCING WELLS 1:7:80	EST. 1980 PRODUCTION $\times 10^3$ B/D	% CHANGE FROM 1980	EST. PROSPECTIVE AREA $\times 10^3$ SQ MI* ON	OFF
ALGERIA	8.20	131.50	5.6	13.36	1000	1000.0	-10.1	635.0	5.2
ANGOLA-CABINDA	1.20	1.05	0.58	1.78	188	150.0	+2.0	20.0	25.2
CAMEROON	0.20	-	0.027	0.227	49	57.0	+68.7	6.0	4.1
CONGO REPUBLIC	0.66	2.00	0.121	0.781	119	56.0	-	27.0	3.5
EGYPT	2.90	2.97	2.183	5.083	484	585.0	+15.6	310.0	14.5
GABON	0.45	0.50	0.825	1.275	242	145.0	-27.0	23.0	17.7
GHANA	0.0063	-	0.001	0.007	3	2.0	-	2.0	8.0
LIBYA	23.00	23.0	13.03	36.03	886	178.0	-13.4	543.0	32.2
MOROCCO	0.000075	-	0.015	0.015	NA	0.2	-	120.0	24.0
NIGERIA	16.70	41.00	8.04	24.74	1451	2100.0	-9.2	157.0	17.8
TUNISIA	1.652	5.6	0.417	2.06	87	100.0	-	63.4	19.5
ZAIRE	0.130	0.05	0.034	0.164	13	22.0	+10.1	27.0	0.4
TOTAL	55.14	208.47	30.87	85.52	4522	5997.0	-9.2	2176.4	172.1

Source: *Oil and Gas Journal*, 29 December 1980. *Grossling (1976) estimates in some cases are far too large and optimistic.

(4) Southern Africa offshore basins, including Algoa fault basin, South Africa
(5) Natal–Transkei–Zululand basin, South Africa
(6) Mozambique basin, Mozambique
(7) Rovuma basin, Mozambique
(8) Majunga basin, Western Madagascar
(9) Morondava basin, Western Madagascar
(10) Tamatave basin, Eastern Madagascar
(11) Malagasy platform, Southern Madagascar
(12) Tanzania coastal and offshore basins, Tanzania
(13) Lamu basin, Kenya
(14) Mandera Lugh basin, Ethiopia and Kenya
(15) Mogadishu coastal basin, Somalia
(16) Somalia or Ogaden basin, Somalia and Ethiopia
(17) Migiurtina or Cotton basin, Somalia
(18) Hafun basin, Somalia
(19) Northern Somalia coastal basin marginal to the Gulf of Aden

Geoscience literature useful for preliminary basin evaluation is complex and widespread. It is written in English, Afrikaans, French, Portuguese and Italian and is compartmentalized because of old colonial and national subdivisions. However, much of the essential literature bearing directly on petroleum exploration, e.g. well log data or seismic information, is kept confidential both by governments and by oil and service companies. The net result is that essential data for evaluation (e.g. seismic coverage, seismic-stratigraphy maps and sections; well logs; maturation studies; deep structural data) are not available for general study except in few cases.

Six Eastern African countries enclose sedimentary basins and in some cases these basins overlap the national boundaries. The total Eastern Africa sedimentary area exceeds 427 770 square miles, while shelf area to the 200 m water depth mark is more than 132 430 square miles. The great Eastern Africa miogeoclinal wedge of sedimentary rocks both onshore and offshore (down to 3 km water depth) occupies about 1 million square miles. Comparative basin areas for the world, North Sea and Eastern Africa - Madagascar are presented in Table 5, and actual sedimentary basin areas for South Africa, Mozambique, Tanzania, Kenya, Somalia and Madagascar are presented in Table 6.

Taking an overview, in all basins, fill (largely sedimentary) exceeds 35 000 ft and includes petroleum source rocks, good reservoir rocks and seals. Sediments, known mainly from surface data, range in age from Ordovician to Early Carboniferous (Cape supergroup basins of South Africa); through Carboniferous to Early Liassic (Karroo supergroup basins of South Africa, Mozambique, Madagascar, Tanzania, Kenya and Somalia); to post-Karroo Early Jurassic-recent basins in all six Eastern African countries. Oil and gas seeps and shows are known in a number of basins but only sub-commercial gas, gas/condensate and minor oil finds have been made.

Table 4
Drilling density and petroleum found per unit prospective area

Region	Prospective area[1] (square mile)	Wells per square mile[2]	Oil EVRD (bbl per square mile)	Gas EVRD (million cu ft per square mile)
Africa (excluding A below)	3 319 310	0.00094	5 578	4.1
Australia and New Zealand	2 440 000	0.0010	2 007	31.1
S and SE Asia mainland	1 558 100	0.0049	6 084	43.8
Algeria + Egypt + Nigeria + Libya = A	1 715 280	0.0057	63 446	242.0
Middle East	1 344 804	0.0059	481 468	842.6
Latin America (excluding B & C below)	3 063 600	0.0068	3 535	16.2
China, P. R.	1 093 000	0.0078	44 282	54.2
Western Pacific	1 617 929	0.0085	20 734	53.8
Argentina + Mexico = B	1 225 000	0.033	29 090	43.8
Western Europe	1 394 000	0.036	50 661	330.2
Canada	1 887 000	0.079	12 617	83.1
Colombia + Trinidad and Tobago + Venezuela = C	516 000	0.082	128 042	71.8
USSR	3 480 000	0.152	66 699	488.3
USA, conterminous	2 534 000	0.957	68 554	387.8

1 Onshore + offshore
2 Total wells (exploratory + development)

EVRD = Expected Value of Recoverable Discoveries i.e. the recoverable petroleum that with practical certainty can be said to have been found so far in a given country consisting of cumulative production and expanded proven reserves (R_2) (Grossling 1976)

Source: Grossling (1976).

Table 5
Comparative areas: world, North Sea 52-62 °N and Norwegian offshore 62-71 °N and Eastern Africa — Madagascar

TOTAL WORLD Shelf & Rise	60 000 000
TOTAL WORLD Shelf	27 000 000
TOTAL EUROPEAN SHELF	1 800 000
TOTAL NORTH SEA SHELF	600 000
BRITISH NORTH SEA 56° - 62°N (within 200m)	48 000
NORWEGIAN NORTH SEA 56° - 62°N (within 200m)	31 000
NORWEGIAN CONTINENTAL MARGIN 62° - 71°N (within 200m)	21 700
EASTERN AFRICA SEDIMENTARY AREA (LAND)	427 777
EASTERN AFRICA SHELF AREA (0-200m)	132 431
MADAGASCAR SEDIMENTARY AREA (LAND)	65 637
MADAGASCAR SHELF AREA (0-200m)	69 498
TOTAL EASTERN AFRICA (LAND + 0-200m) + MADAGASCAR (LAND + 0-200m)	695 343

Post-Karroo (largely Mesozoic and Cenozoic) sedimentary rocks in these basins are more or less continuous onshore and may be viewed as the onshore portions of the great Eastern Africa miogeocline which extends from South Africa to Somalia, along what is now the western margin of the Indian Ocean. The sedimentary rocks constituting the Eastern African miogeocline and the Gulf of Aden protomiogeocline (Somalia also has a shore on the Gulf of Aden) thin both landward and oceanward and were formed as a result of continental margin subsidence related to thinning of what is now the eastern margin African lithospheric plate, known as Pangaea. The Palaeozoic super-continent broke up into Eastern and Western Gondwanaland in Early Mesozoic time and later. This plate history is complex and is not well known. It is relevant to evaluation studies in that it provides the broad backdrop against which detailed specific basin evaluation can be started.

Beneath the Late Mesozoic-Cenozoic rocks of the Eastern African miogeocline, and in places on the landward flank of the sediment wedge, from Somalia in the north to Mozambique in the south, there is evidence of an extensive and more or less continuous Karroo rift trough system. This has been explored to a limited extent only. Few published geological maps of Africa show the true extent of this trough system, which extends both onshore and offshore, and the Tectonic Map of Africa (1 : 5 000 000 scale 1968) only shows a small part of it. Karroo rift troughs have been identified in the onshore areas of Mozambique, Madagascar, Tanzania, Kenya and Somalia, where the sedimentary fill is tens of thousands of feet thick and where the troughs form

Table 6
Sedimentary areas onshore and offshore and total land areas: South Africa, Mozambique, Tanzania, Kenya, Somalia and Madagascar

	South Africa		Lesotho		Madagascar		Mozambique		Tanzania		Kenya		Somalia	
	sq km	sq mi	sq km	sq mi	sq km	sq mi	sq km	sq mi	sq km	sq mi	sq km	sq mi	sq km	sq mi
Total Land area	(1 112 000)	472 200	(30 344)	11 715	(590 000)	227 799	(784 960)	303 073	(939 800)	362 856	(582 600)	224 941	(637 000)	245 945
Main Sedimentary area (land)	(260 000)	100 385	(17 000)	6 563	(170 000)	65 637	(187 000)	72 200	(52 000)	20 077	(192 00Q)	74 112	(400 000)	154 440
Shelf Area (0 - 200 m)	(143 000)	55 212	-	-	(180 000)	69 498	(105 000)	40 540	(21 000)	8 108	(14 000)	5 405	(60 000)	23 166
Deeper Water Area (200 - 3000 m)	(487 000)	188 030	-	-	(270 000)	104 247	(160 000)	61 776	(100 000)	38 610	(60 000)	23 166	(300 000)	115 830

part of an extensive 'failed rift' system which developed in Late Palaeozoic and Early Mesozoic times. Such rift systems are known generally as favourable environments in which oil and gas may have accumulated.

In the offshore of Madagascar, Tanzania, Kenya and Somalia, Karroo troughs have been recognized but they are less conspicuous and much less is known about them compared with the onshore occurrences, largely because later Mesozoic and Cenozoic deposits and subsidence have greatly modified and buried the trough systems. Karroo evaporites have been identified in Tanzania, both at the surface and in the subsurface, but are thought to occur elsewhere (Figs 9 and 11). Late Mesozoic and Cenozoic evaporites, of different provenance from the Karroo, are known from Somalia.

Over the last 100 years in South Africa, and over the last 50 years in Madagascar, Mozambique, Tanzania, Kenya and Somalia, explorationists have prospected the Eastern Africa basins but without commercial success. Prospecting has proceeded at various rates, been undertaken by a range of organizations and conducted at a variety of levels; it is continuing in all countries at the present time but it is proceeding at a slower rate except in South Africa. It was only in South Africa and Tanzania that wells were drilled in 1979. In all six of these countries, excluding South Africa, exploration has been mainly onshore. Only 26 wells have been drilled in the offshore up to the end of 1980. The limited commercial success after 35 years of post-World War-II search poses important questions for the countries concerned, which all have fundamental economic, financial and political problems of varying degrees of severity deriving largely from having to import petroleum. These include the following:

(1) Should the lack of commercial success onshore be taken to indicate poor prospects for exploration in the shallow and deep offshores and will the outer shelves and upper continental slopes, if they are explored, be rewarding enough to justify exploration and development costs? What kind of geology underlies these areas?
(2) Have these basins been adequately explored over the last 30 years or so, or are we assessing these basins in terms of dry holes located on structures identified 'dimly' using poor quality or technically outmoded seismic data, etc.?
(3) Has fate been cruel enough to locate what petroleum prospects the countries can lay claim to in the Karroo troughs and elsewhere offshore where the ocean is too deep currently for economic development; or to locate hydrocarbon resources in difficult-to-exploit heavy oils or 'tar sands' as in Madagascar?
(4) Have adequate portions of the stratigraphic columns attained high enough threshold temperatures for oil and gas to have been generated from source rocks? Other pull-apart miogeoclinal margins such as the eastern seaboard of the United States and Canada have poor success records.
(5) Can these basins, which occupy around 1 million square miles (to

Table 7
Exploration wells drilled, successes, and footage drilled 1949-1979, Eastern Africa

	PERIOD COVERED	EXPLORATION WELLS	SUCCESSFUL	TOTAL	FOOTAGE DRILLED
SOMALIA	1949 - 1974*	48	0	48	429 785
	1975 - 1979	3	0	3	32 271
	1949 - 1979	51	0	51	462 056
KENYA	1949 - 1974*	13	0	13	98 462
	1975 - 1979	4	0	4	45 183
	1949 - 1979	17	0	17	143 645
TANZANIA	1949 - 1974*	14	0	14	108 369
	1975 - 1979	5	2 (GAS)	5	38 878
	1949 - 1979	19	2	19	147 247
MOZAMBIQUE	1949 - 1974*	51	4	51	488 523
	1975 - 1979	0	0	0	0
	1949 - 1979	51	0	51	488 523
MADAGASCAR	1949 - 1974*	52 (58)	0	52 (58)	422 303
	1975 - 1979	0	0	0	0
	1949 - 1979	52 (58)	0	52 (58)	422 303
SOUTH AFRICA	? - 1975*	103	2	114	662 304
	1975 - 1979	48	0	48	365 671
	- - 1979	151	2	162	1 027 975
TOTAL WELLS DRILLED	1975 - 1979	60 (66)	2	60	
TOTAL					3 903 151

Source: Grossling (1976) data up to 1975. Additional 1975-1979 data from: 'Foreign Developments', *American Association of Petroleum Geologists Bulletin*. Anon (1980), Reserves, *Oil Gas J.*, 29 December 1980.

3 km water depth) or two-thirds of a million square miles (to 200 m water depth) extending over a straight line distance of 5000 miles, characterized by favourable and diverse geology and in some cases possessing positive hydrocarbon indices, be devoid of commercially viable accumulations?

These are some of the questions which come to mind in trying to build up thumbnail sketches of the petroleum geology and prospects and in speculating about petroleum resources for these six Eastern African states.

Answers to some of the questions can be given straight off. Sedimentary rocks in some of the basins are known to have low gaseous and liquid hydrocarbon potential — for instance, the sedimentary formations of the Cape and Karroo supergroup basins of the Cape fold belt, South Africa, and the Karroo and younger formations of the onshore rift troughs of Tanzania. Many explorationists have written off the Cape and Karroo basins of South Africa because it is unlikely, for good geological reasons, that the Cape fold system, and most of the Karroo basins and platform will yield large accumulations of oil and gas, and it is reasonably certain that large areas are unlikely to yield even small accumulations. The Tanzania Karroo troughs are considered by some geologists to have been adequately tested and to be devoid of sizeable accumulations.

Even small quantities of oil or gas would be welcome in all these Eastern African countries bordering on the Indian Ocean. The discovery of larger quantities would contribute perhaps to the development of adjacent inland states and to the improvement of the world energy situation as a whole. Certainly recent political events have retarded exploration and development and thorough investigation of prospects in five of these countries; but not so in South Africa, where the laager syndrome has driven SOEKOR (South Africa's national oil company) to drill extensively both onshore and offshore. Only in the Agulhas Bank area have encouraging signs been found, however.

Fewer than 310 wells had been drilled in these basins from the Cape of Good Hope to Cape Gardafui up to 1975. Since then up to the end of 1979 only 60 more wells have been drilled, and only two of these 1975–1979 wells are rated as successes (Songo Songo area, offshore Tanzania). Footage drilled in the six Eastern African countries 1949–1979 is 3 903 151 ft, 26% of it in South Africa (Table 7).

PART II
SUMMARY OF PETROLEUM GEOLOGY AND PROSPECTS

SOUTH AFRICA

Sedimentary rocks occupy about one-quarter of the surface area (472 000 square miles) of the Republic of South Africa, Lesotho (11 715 square miles) and Swaziland (6705 square miles). For more than 100 years minor shows and occurrences of oil and gas and related semi-solid and solid hydrocarbon substances, in all totalling over 350, have been recorded mainly from the Karroo supergroup both from the surface and the subsurface. These led prospectors on, and over the years a large number of wells were drilled in trying to locate 'seep' oil in commercial quantities.

The sedimentary outcrop area is roughly kidney-shaped and is more than 850 miles long and 370 miles wide. Stratigraphic units for South Africa are listed in Table 8. Within this vast area the Cape and Karroo supergroups both attain cumulative thicknesses of 33 000 ft in the southern part of the area, where they are involved in folding, thrusting, low grade metamorphism and high grade diagenesis. In addition to these great thicknesses of Cape and Karroo sediments, small basins exist, both on land and offshore, and contain large thicknesses of Mesozoic (mainly Cretaceous) and Cenozoic rocks. In the Algoa Basin sediments exceed 18 000 ft. In the large offshore basin developed north of Superior Oil Company's small gas find south-east of Plettenberg Bay, sedimentary rocks exceed 24 000 ft.

Onshore South Africa for many years has appeared a paradox to oil seekers being a country with extensive sedimentary basins containing

Table 8
General classification of stratigraphic units, South Africa, used in this report

STRATIGRAPHICAL UNITS		AGES
Various localised formational units		TERTIARY - QUATERNARY
UITENHAGE GROUP and various formations		CRETACEOUS
KARROO SUPER-GROUP	STORMBERG GROUP	TRIASSIC - JURASSIC
	BEAUFORT GROUP	PERMIAN - EARLY JURASSIC
	ECCA GROUP	PERMIAN
	DWYKA GROUP	CARBONIFEROUS - PERMIAN
CAPE SUPER-GROUP	WITTEBERG GROUP	LOWER CARBONIFEROUS - DEVONIAN
	BOKKEVELD GROUP	DEVONIAN
	TABLE MOUNTAIN GROUP	ORDOVICIAN - SILURIAN
KLIPHEUWEL FORMATION		? CAMBRIAN
NAMA AND DAMARA GROUPS		LATE PRECAMBRIAN - CAMBRIAN
----UNCONFORMITY----		
PREDOMINANTLY CRYSTALLINE AND METAMORPHIC ROCKS		PRECAMBRIAN

suitable reservoir rocks of Palaeozoic, Mesozoic and Cenozoic ages; with structures and numerous 'shows' of oil, gas and related substances; and with possible source materials but without commercial oil and gas finds.

Table Mountain sandstone, Bokkeveld and Witteberg groups of Cape Province Palaeozoic supergroup basin

In reality the so-called Table Mountain sandstone (Ordovician to Lower Carboniferous) is not a sandstone but a metaquartzite and in many areas has attained a low grade green schist facies of metamorphism. The so-called Bokkeveld shales are in many places *bona fide* slates and the Bokkeveld and Witteberg sandstones are largely quartzites. All the groups clearly show the effects of deep burial and advanced diagenesis. Porosities in the Table Mountain group are low and range from 0 to 8%. Hydrocarbon potential and prospects are rated as very low to nil as a consequence of high diagenesis and low metamorphism. Montmorillonite has been transformed largely to illite in shales and other lithological units in large sections of the geological column. Hydrocarbons could only have accumulated in such rocks in areas where secondary fracture porosity and permeability have developed.

The main reasons why vast volumes of folded Palaeozoic sedimentary rocks are barren of hydrocarbons in South Africa is that the timing of generation of oil and gas, migration, development of structure, generation of secondary porosity and fracture porosity did not coincide over large areas, so leaving 'dry' many of the great and spectacular structures of the Cape Fold Belt.

Karroo supergroup basin and platform prospects

Combined maximum sediment thicknesses exceed 33 000 ft in the Karroo supergroup basin and on the platform. The Karroo sediments in part were involved in the Cape Orogenic Episode (Triassic?) but are tectonized and metamorphosed on a smaller scale along the northern margin of the Cape Fold belt and in the Cape syncline than the Cape supergroup. Karroo sediments on the Karroo platform are much thinner than in the Cape syncline but show the effects of deep burial. They are gently dipping to flat lying and are extensively intruded with dolerite dykes and sills (Figs 3 and 4). Hydrocarbon potential is rated as low despite many shows of gas and oil. These are thought to have been formed by thermal distillation of Ecca 'oil shales' and related sediments and Ecca coals. Small finds of gas can be expected in structurally suitable areas.

Dwyka group prospects

Porosities and permeabilities of the Dwyka group are low and reservoir potential is low also. Again fracture porosity and permeability have developed in places but apparently not on a regional scale suitable for oil or gas accumulation. The Upper Dwyka shales are said to have source rock characteristics.

Ecca group prospects

Porosities and permeabilities of around 15% and 1 Darcy are known from some middle Ecca sandstones in the northern outcrop area of the Karroo supergroup but elsewhere Ecca porosities are low and fracture porosity is localized. Sand and silts are uniformly tight over large areas. Some Ecca mudstones have source rock characteristics and Ecca coals must be considered potential sources for 'dry' gas. Prospects for oil are considered to be low over-all but possibilities may exist for small local accumulations on the northern Karroo platform.

Beaufort group

Rocks with source and reservoir rock characteristics are not common in the Beaufort group and dolerite, sills, sheets, etc. riddle the formation (Fig. 3). Dissection and lack of cover affect the group's hydrocarbon holding capacity. Prospects are very poor.

Stormberg group prospects

Stormberg outcrops are much less widespread than the Beaufort group and the extrusive formations are developed mainly in Lesotho. The Drakensberg volcanics of Early Jurassic age are as much as 4500 ft thick

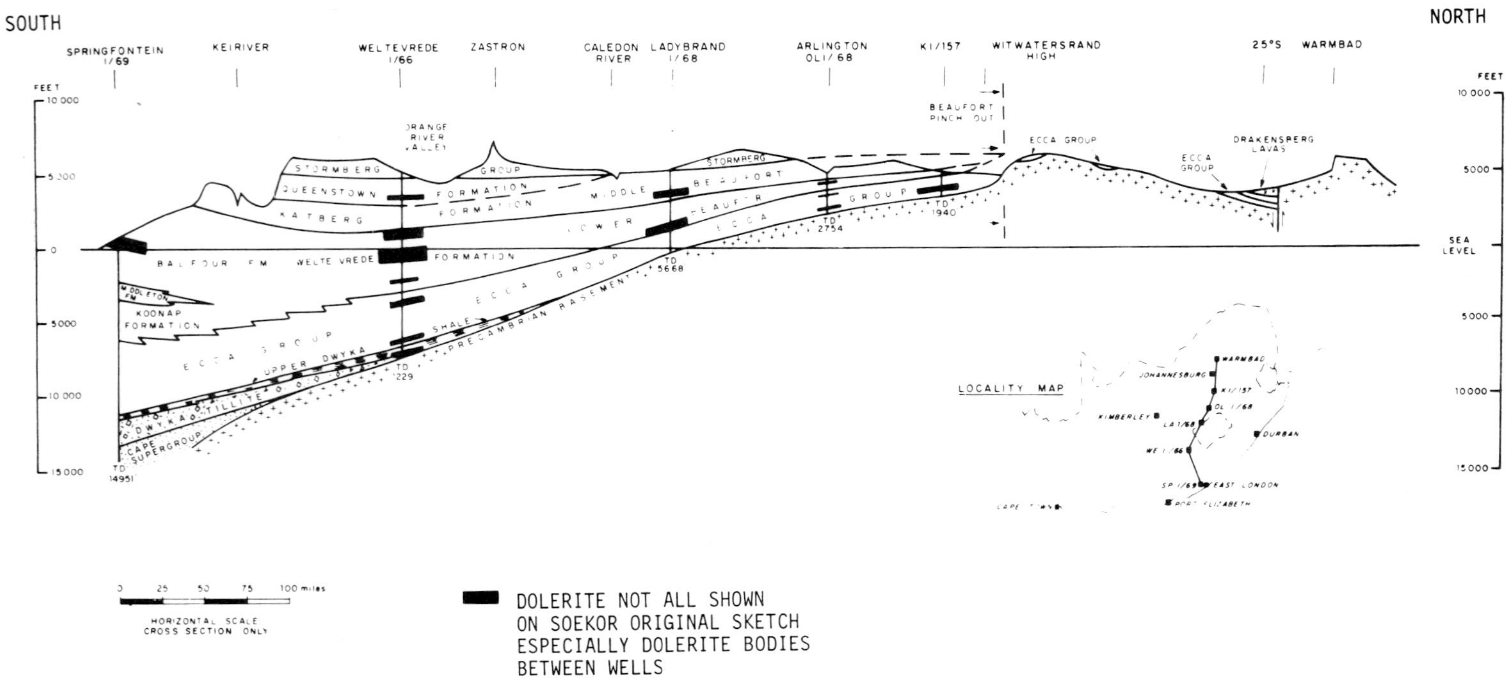

Fig. 3. Regional cross section, eastern Karroo basin (and Witwatersrand High). (Source: Soekor, 1971).

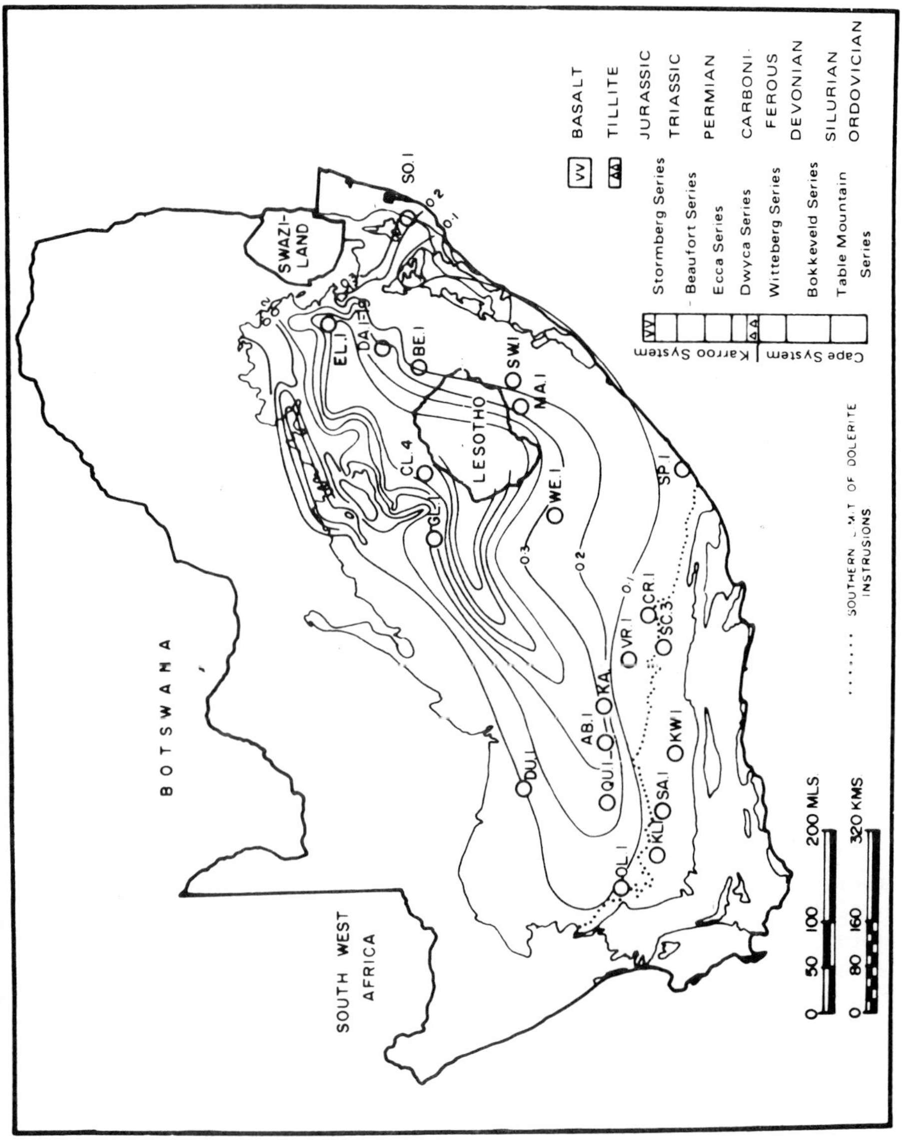

Fig. 4. Map showing dolorite sediment ratio for South Africa and Lesotho. Based on Winter and Venter (1970).

in places and cap the Karroo sequence. Karroo dolerites are a widespread hypabyssal phase of the Drakensberg volcanics; they intrude the Karroo sediments extensively and indicate that the volcanics must have once been much more widespread.

Rocks with source rock characteristics are not developed in the Stormberg sediments which underlie the Drakensberg volcanics but good reservoir rocks occur in the Molteno beds. The Drakensberg volcanics are regarded by some as an impervious cover and plans for prospecting for gas in Lesotho beneath the lava cover have been mooted. Over-all, the prospects of the Stormberg sediments are poor because of dolerite intrusion; high palaeo-geothermal gradients which may well have over-matured sediments; and deep dissection during the Cretaceous and Cenozoic periods.

Dolerite dykes, sheets, etc.

Although not shown in the 1 000 000 scale geological map of South Africa (1970), dolerite dykes, sheets, sills, etc. are extensive north of the Great Escarpment or north of an east–west line through Beaufort West and Aberdeen. Many of the 'seeps', 'shows', semi-solid occurrences, etc. of hydrocarbons are related to small-scale thermal distillation of the Ecca shales by dolerite intrusions. As a consequence 'seepology' as practised in the early days of exploration in South Africa was doomed to failure because of the relationship of many seeps to dolerite intrusions. This situation is unlike that prevailing in parts of the United States, where seeps are frequently related to anticlines and consequently the practice of seepology led to many commercial successes.

Intermontane Mesozoic basins, Cape fold system: largely Cretaceous and Cenozoic prospects

Fifteen small fault basins containing Mesozoic and Tertiary clastic sediments occur with the Cape fold system in South Africa (Fig. 5). Sediment thicknesses in these basins range from around 1000 ft to over 15 000 ft. The basins are flap fault basins in the main with an estimated structural throw of over 12 000 feet known in the case of the Worcester fault. The Algoa fault basin, which continues into the offshore, had minor gas shows in wells, and in the offshore gas was tested in Placid's F-1 sub-commercial well south of Mossel Bay and in Superior's gas find situated south of Plettenberg bay. Gas in G (a) A-1 well was produced from the Cretaceous at 25 MMCFD with about 100 B/D condensate and at 10 MMCFD from around 7200 ft from fractured Table Mountain sandstone. These basins, especially in the offshore, may have considerable exploration potential. Between 1977 and 1979 at least six wells were drilled on the Agulhas Bank area which had either minor uneconomic gas or oil shows; and in October 1980 SOEKOR's ED-4 was said to be testing gas. The political situation, however, places severe constraints on exploration, except mainly by domestic companies. The potential of the onshore basins is thought by many to be low or negligible, simply

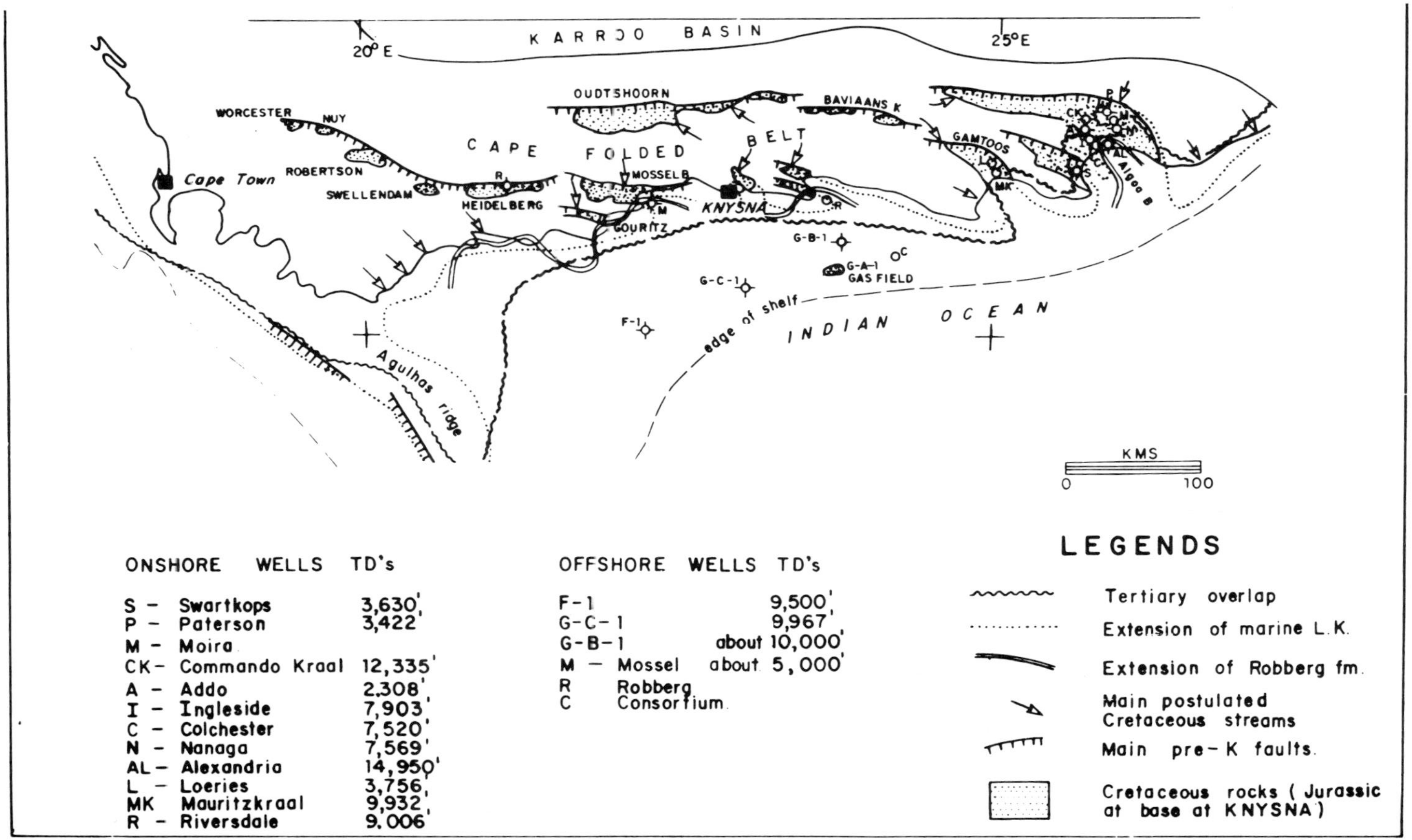

Fig. 5. Mesozoic formations: map showing distribution of Cretaceous rocks and localities, Cape Province to Algoa Bay (Rigassi and Dixon, 1972).

because of the lack of suitable source rocks and an inadequate maturation history. Nevertheless, wildcatters still prospect some of these Cape-Karroo basins. Up to 1975 some 80 wells had been drilled specifically for oil and gas in South Africa but only 17 wells had been drilled in the offshore. At the present time, more than 37 holes have been drilled in the offshore in the Agulhas Bank area; none have been drilled in the north eastern offshore, except JC-1 north east of Durban.

Natal, Cape supergroup Palaeozoic basin

The fragmented western portion of the Natal Cape supergroup basin extends northwards from Port St. John's to Hluhuwe Game Park a few miles west of the south end of the Lebombo monocline. Only the Table Mountain sandstone group of the Cape supergroup is preserved in Natal, where it attains a maximum thickness of around 4000 ft north east of East London. How much of the Cape supergroup is present offshore beneath the Karroo and other Mesozoic and Cenozoic strata within accessible production depths is unknown publicly. Because of the diagenetic history of the Cape supergroup elsewhere in South Africa the hydrocarbon prospects of the supergroup in onshore and offshore Natal are considered to be low by many explorationists. However, sporadic gas pockets (and oil?) may occur in suitable fractured rocks if we use the Aghulas Bank situation for analogy.

Transkei - Zululand - Mozambique, Mesozoic-Cenozoic miogeocline

This great sedimentary wedge, which developed over the western flank of a Karroo failed arm, formed as Africa, Australia, Antarctica and Madagascar spread apart. It consists of Mesozoic (predominantly post-Jurassic) and Cenozoic deposits which at the present shore line exceed 35 000 ft. Karroo sediments and volcanics occur on the western margin of the miogeocline and in the adjoining Limpopo and Zambesi – Shire failed rift troughs. Karroo and Cretaceous fault troughs occur beneath the coastal area and offshore. The hydrocarbon prospects of the miogeocline are poorly known and we are dependent on the Deep Sea Drilling Project and the scientific survey ships for basic geological information.

MOZAMBIQUE:
SUMMARY OF THE PETROLEUM GEOLOGY AND PROSPECTS

Exploration for oil and gas began in the early 1900s, but only minor commercial production has been attained. Two major Mesozoic-Cenozoic sedimentary basins occur in Mozambique and these developed over faulted and down-warped Karroo rocks (Carboniferous–Early Triassic). As in adjacent areas in Tanzania, Zimbabwe and Zambia, Karroo rocks are disposed in one or two sided fault or monocline bounded troughs.

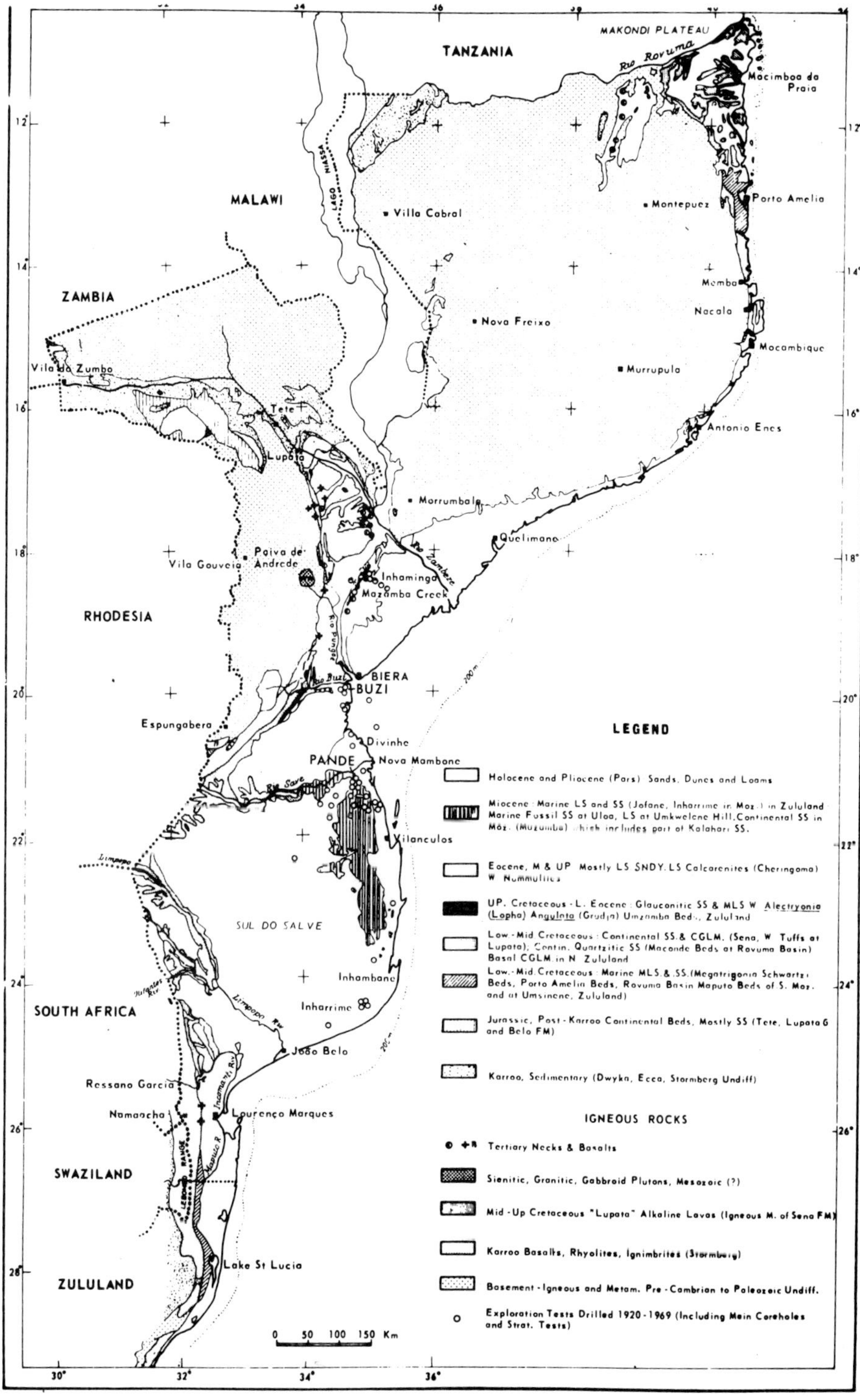

Fig. 6. Geological sketch showing the onshore Mesozoic and Tertiary basins of Mozambique and Zululand. Based on Flores (1970).

The Mozambique basin is the largest and southernmost of the coastal basins and extends from Zululand (South Africa) where the continental shelf widens to a point some 400 miles north east of where the Zambesi River enters the sea. The basin is 250 miles wide maximum onshore but extends into deep water, forming part of the Eastern African miogeocline. The smaller onshore–offshore Rovuma basin extends into Tanzania, where gas has been found at Songo Songo. Both basins are characterized by vertical tensional tectonics produced as the Mozambique Channel rift system evolved. Compressional folds are unknown in the Karroo supergroup and post-Karroo rocks; drape folds related to fault blocks are known. The Mozambique basin connects with the Lower Zambesi failed rift trough and the Shire failed rift trough. Coals in the Karroo supergroup could have been a source for gas in these areas.

There has been no drilling in the Mozambique section of the Rovuma basin; in contrast, up to 1976 some 58 wells had been drilled in the Mozambique basin (excluding shallow holes). Only the Pande field, 100 miles south of Beira, discovered in 1961–1962, is a potential gas producer. The production rate of Pande-1 was estimated at 4.5 MMCFD through a 3/8 inch choke. Production was from the Grudja sand at about 3750 ft. Pande-4 (Gulf and Panamoz) blew out and Pande-4D relief well was abandoned in January 1967. The Pande field is now shut down. Estimated recoverable reserves are put at 0.1×10^{12} cu. ft. Buzi-1, drilled in 1962 25 miles south west of Beira, had encouraging shows of gas and condensate. No data appear to be available publicly about the size of the find. Buzi-1 was completed as a non-commercial gas discovery. As in Pande-1, the gas-bearing reservoir is in the Upper Cretaceous strata.

The greater part of the coastal basin onshore is reasonably well known geologically in terms of surface geology, seismic coverage, well data, etc. The south western part of the Mozambique basin is poorly known, however. Negative results of wells drilled in the late 1950s and early 1960s may have diverted attention from the coastal areas, and little further investigation has been carried out since then. The coastal area is characterized by small traps and although Cretaceous source and reservoir rocks are widespread, the onshore area is considered to have poor potential. However, Karroo troughs may have some potential for gas derived from Karroo coals but such prospects are considered highly speculative onshore and almost certainly out of current economic drilling depth offshore. Viewed over-all, onshore Mozambique is thought to have only poor to fair potential by some geologists because enough wells have been drilled onshore, and in the shallow offshore, to get a 'feel' for the currently accessible prospects. This may be a narrow view, however, especially if the prospects are viewed in national rather than international terms.

Concerning the offshore prospects formerly held by Sun, Amoco and Elf Aquitaine, these are considered to be speculative also. However, Cretaceous and Tertiary clastics with marine shale sources, and sealed in by such shales, associated with horst block structures developed over the basement complex and Karroo formations, could be accessible targets beneath the shallow and deep parts of the shelf. In addition there might well be plays in the Zambesi, Save and Limpopo delta complexes. Large basins are known to underlie the inner and outer shelf areas and the continental slope. Little is known publicly about the stratigraphy and structure of these delta complexes, and even less about geothermal gradients and the maturation levels prevailing therein. The Palaeogene–Neogene delta complex sequences are thin, compared with the Niger and Nile delta sequences, for example, although depth to basement complex at the coast exceeds 24 000 ft in the Zambesi delta.

The gas-prone nature of the upper part of the deltaic and Cretaceous sequences proven so far has yet to be explained adequately. Mozambique might be a gas province and if high geothermal gradients are associated with the Lower Zambesi rift trough, where there are Karroo Cretaceous and Cenozoic volcanics, then these might increase the tendency of deltaic gas-prone source materials to generate gas. No geothermal or source and maturation data have been published for Mozambique.

Concerning the deep water prospects of the Mesozoic – Cenozoic continental margin wedge and block faulted Karroo formations which occur in younger sediments, magnetic and seismic information indicates thicknesses of over 30 000 ft. Some sources consider about half this section to be prospective, especially the Early Cretaceous to Mio-Pliocene portions. Plays can be envisaged in Lower and Upper Cretaceous and Tertiary sands interfingering with, sourced by and sealed in by shales: block faults, continental margin subsidence structures, drape overs, mud diapiric bodies (?), channel sand plays and carbonate build-ups might occur. As we have noted, the coastal portions of the onshore basins have been reasonably well explored in sections but only 11 wells have been drilled in the offshore areas. The shallow water area is considered to have been inadequately explored by some geologists; the deep water area will only be explored when an adequate rate of return can be guaranteed because only large fields can be viable in such areas.

The general view about Mozambique's prospects is that there may well be enough oil and gas awaiting discovery to satisfy internal needs, but because of poor results in the past, recent political and economic uncertainties, etc., few companies are willing to explore further. No drilling has taken place in Mozambique since 1975. No petroleum rights are currently held; BRASPETRO is said to be currently investigating petroleum potential and AQUATER (ENI) to be planning a geochemical survey (29 000 km^2) in the Zambesi and Nampula region and in another unspecified area (19 000 km^2). Field geological surveys are also planned.

WESTERN MADAGASCAR MIOGEOCLINE

During Karroo (Carboniferous to Early Jurassic) times fault bounded basins developed consequent on lithospheric plate thinning localized between the African mainland (Transkei – Zululand – Mozambique – Tanzania) and Madagascar.

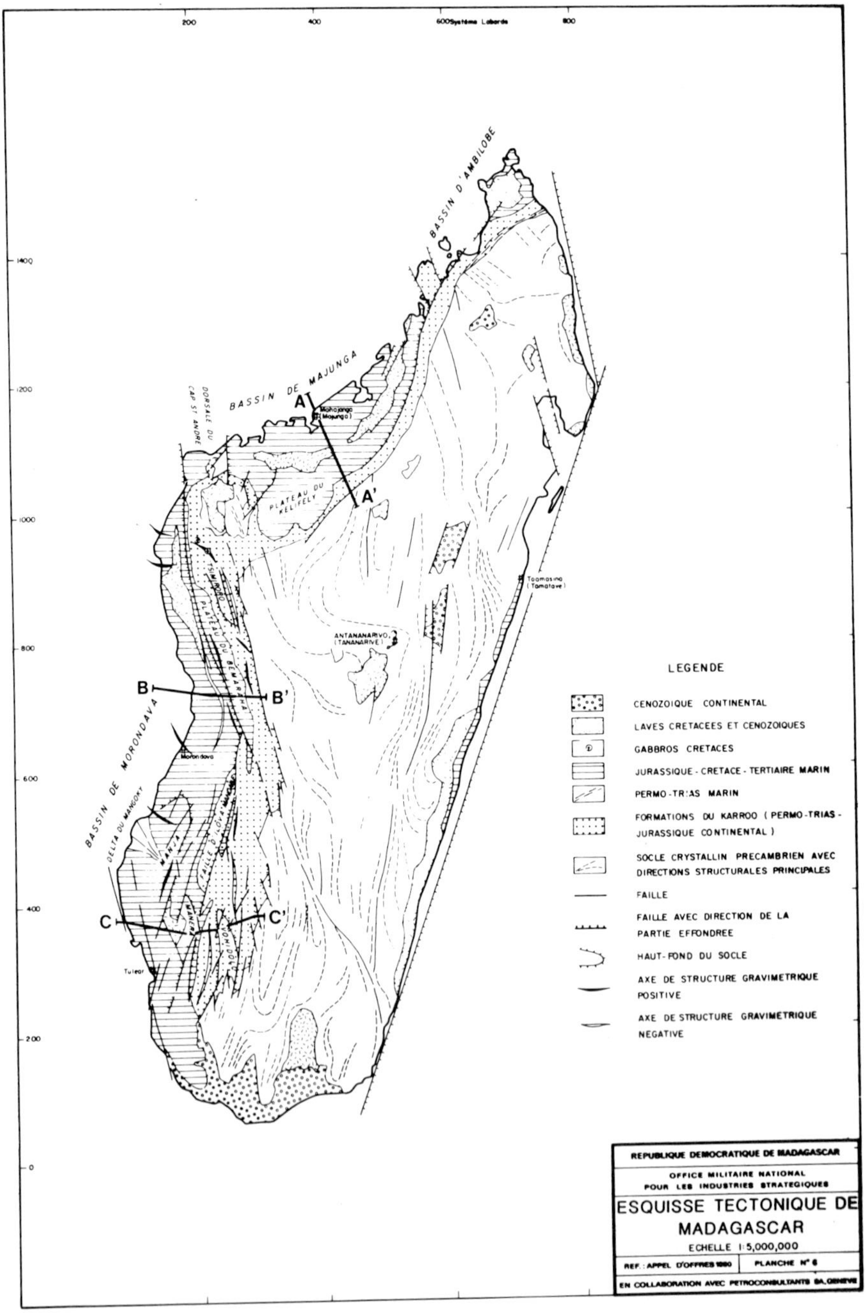

Fig. 7. Tectonic sketch of Madagascar.

After separation was accomplished, and new crust had been generated beneath what is now the Mozambique Channel, the early stage rift basins were modified and buried as the continental margin subsided in Jurassic, Cretaceous and Cenozoic times. Two main basins occur onshore in Madagascar: the south western or Morondava Basin and the north western Majunga Basin. In both areas the Karroo supergroup rests directly and disconformably on the basement complex and is overlain by the remainder of the Mesozoic and by Cenozoic deposits. Surface petroleum seepages are abundant and there are extensive tar sand deposits in the Demolonga area. Nevertheless, despite the fact that many wells have been drilled, some with oil and gas shows, commercial accumulations have not been found to date (Fig. 7).

As on adjacent parts of the mainland, the search for petroleum in Madagascar started early and by 1951 wells had been drilled in the vicinity of hydrocarbon shows in the Tulear area. Over the preceding 50 years considerable amounts of geological and geophysical data had accumulated.

The Morondava and Majunga basins have been mapped and described in great detail. It is estimated that Bemolonga tar sands (surface and near surface) contain reserves of *ca.* 3×10^9 tonnes (i.e. equivalent to 22×10^9 bbl) of tar in place.

Up to 1960 the Société des Pétroles de Madagascar (SPM) had drilled about 40 exploratory wells in the Morondava basin. From 1961 to 1967 exploration was concentrated in the Majunga basin, where two wells were drilled. From 1968 to 1975 a number of new companies engaged in exploration both offshore and onshore and some 15 wells were drilled. However, no commercial finds were made.

From 1976 the Office Militaire National pour les Industries Strategiques (OMNIS) became responsible for exploration for hydrocarbons in the Democratic Republic of Madagascar. Marine and onshore geophysical surveys were carried out on behalf of OMNIS and recently invitations to submit bids were issued with the aim of promoting exploration for hydrocarbons in both onshore and offshore Madagascar.

Both the Morondava and the Majunga basins contain continental and marine sedimentary rocks of Permo-Carboniferous (Karroo) to recent age. Representative sections showing the stratigraphy and structure of the Majunga and Morondava basins are presented in Fig. 8.

Potential source rocks such as the Sakoa coal (gas) belonging to the Upper Carboniferous - Lower Permian rocks, and black and grey shales (oil and gas) rich in plant matter (Permian, Triassic, Jurassic and Lower Cretaceous) occur. Although source data are not conclusive, most of the heavy oils and tars are considered to have been generated by the Permo-Triassic Sakamena formation.

Geothermal gradients are reported to be high in both basins (average 3.5 °C/100 m, i.e. 2 °F/100 ft). Large volumes of source rocks appear to have been matured and some parts of the column are known to have been over-matured. The plant-rich source may indicate that the basins will prove to be gas prone. Structural and stratigraphic traps are abundant.

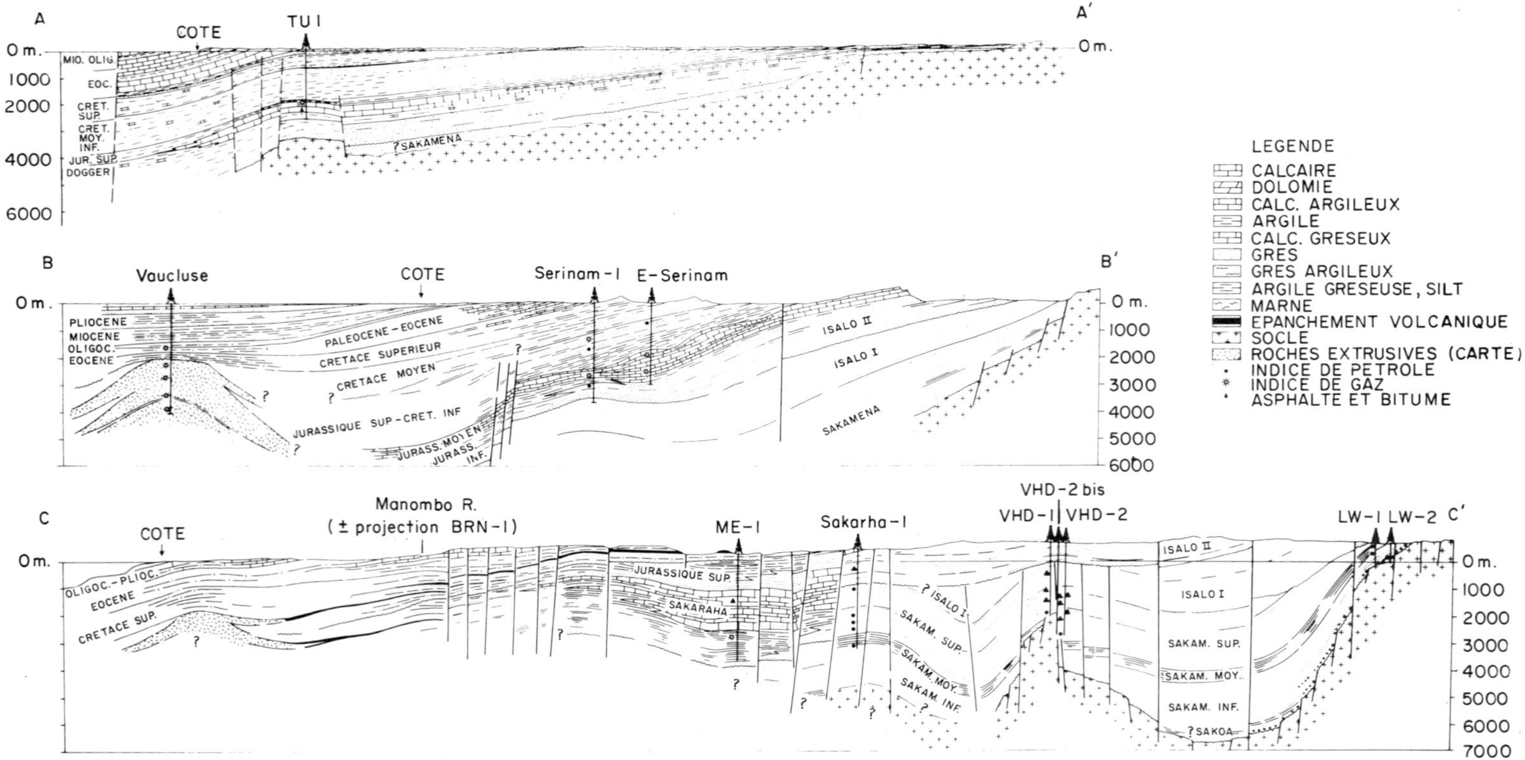

Fig. 8. Structural sections: Majunga and Morondava basins, Madagascar (OMNIS-Petroconsultants, 1980).

Compressional structures are unknown and tensional tectonics predominate. Several unconformities have been recognized, the most striking one occurring near the base of the Cretaceous.

Potential reservoir sands occur in all formations from the Permian to the Palaeogene. Carbonate reservoir rocks occur in Jurassic–Tertiary rocks. Adequate shale seals exist in general terms. Some reservoirs appear to be meteoric water flushed.

Despite the many positive hydrocarbon indications in the Morondava and Majunga basins, exploration results so far have been disappointing. Some of the 56 exploratory wells tested structures which were not closed and some of the early wells drilled in the 1960s were not drilled deep enough to test all possible accumulations. OMNIS and its geological advisers (1980) think that sizeable accumulations of light oil and/or gas must occur within the two basins and that once conditions of generation, migration, entrapment and retention are properly understood, and additional exploration undertaken, there is a good chance that commercial successes will be achieved.

OMNIS (1980) highlighted four zones which they considered to hold considerable prospective interest:

(1) Manja (9600 km^2) onshore coastal area in the south central Morondava Basin;
(2) Belo-sur-Mer (9760 km^2) onshore and offshore in the central Morondava Basin;
(3) Sud Serinam (6250 km^2) east of Belo-sur-Mer in the central Morondava Basin;
(4) Majunga offshore (14 160 km^2) in the Majunga Basin.

Extensive seismic coverage exists for these areas and principal objectives are Eocene and Upper Cretaceous sandstones and carbonates involved in dip closed structures.

Very little information is available on Madagascar's other basins situated offshore on the eastern and southern sides of the island. Along the east coast, sediments are lying at 2000 m water depth, on the down side of the major fault shaping that part of the island.

TANZANIA

Attention was drawn to the hydrocarbon potential of the Tanzania coastal basins by the Tanganyika Geological Survey and BP-Shell. With BP as operator in East Africa, the group embarked in 1952 on an exploration programme for oil and gas. Coastal sedimentary basins and the offshore islands were included in these investigations.

Some 52 exploratory wells were sunk as a result of this programme and a great deal of geophysical work was completed both onshore and offshore, records of which are filed with the Geological Survey of Tanzania and the Tanzanian Petroleum Development Corporation.

Extensive gravimetric surveys (gravity and magnetic) were made. No commercial oil or gas finds were discovered, however, and very few shows were reported. BP-Shell eventually abandoned the venture in the early 1960s. BP geologists and geophysicists provided a systematic account entitled 'The Geology and Geophysics of Coastal Tanzania' (Kent *et al.*, 1971). Kent (1965) described the geology of the Mandawa salt basin and Kent and Perry (1973) gave a general account of the development of the Tanzanian section of the Indian Ocean margin.

These surveys showed that, east of the concealed Tanga fault belt and elsewhere in the coastal belt, sediments thicken to over 20 000 ft. Gravimetric surveys helped in elaborating basin studies and from 1955 to 1964 five refraction and seven reflection surveys outlined the subsurface geology of the areas around the Pugu, Msanga and the Wingayongo-Ruaruke areas. A marine reconnaissance survey was conducted around Pemba Island in 1959 and later off Zanzibar and Mafia Island.

After BP-Shell abandoned the area, no exploration took place until 1969. In these years no activity was reported except for what the Ministry of Commerce and Industry described as 'continuing inquiries from overseas oil interests'.

In 1969 the Government of Tanzania set up a Government Authority which later developed into the Tanzanian Petroleum Development Corporation (TPDC) in 1974. This authority signed an agreement with the Ente Nazionale Idrocarburi (ENI), the Italian state oil company. AGIP (an ENI affiliate) acted as the main contractor and field work was resumed in 1969. Both geological and marine seismic activities were started and AGIP was awarded the whole concession out to 200 m water depth.

The Tanzanian Petroleum Development Corporation is the sole concessionaire in Tanzania and the contract with AGIP begun in April 1969 is of the service contract type. It covered 19 000 square miles and included the whole of the continental shelf and the islands of Pemba, Zanzibar, Latham and Mafia. The contract was valid for 30 years. All exploration expenses were borne by AGIP.

In mid-1969 the Tanzanian Government signed a technical assistance agreement with the Soviet body Techno-Export for its oil exploration programmes. Early in 1970 an 18-month seismic survey contract was signed between Rogers Geophysical and AGIP. No exploratory wells were drilled in 1970 and only three crew months of seismic work were done. AGIP's acreage position was maintained in 1972. No wells were drilled and AGIP completed 20 party months of seismic survey. In 1973 AMOCO International acquired a 50% interest in the AGIP onshore and offshore concession. AGIP completed three party months of land seismic surveying and spudded the first offshore wildcat, Ras Machiusi-1, on the mainland side of the Zanzibar Channel.

In 1974, AGIP-AMOCO abandoned two dry holes, Ras Machisisi-1 and Songo-Songo-1. Songo-Songo-1 had found gas shows. No other activity took place in 1974.

In 1975 the Oceanic Exploration Company was awarded 40 000 km^2

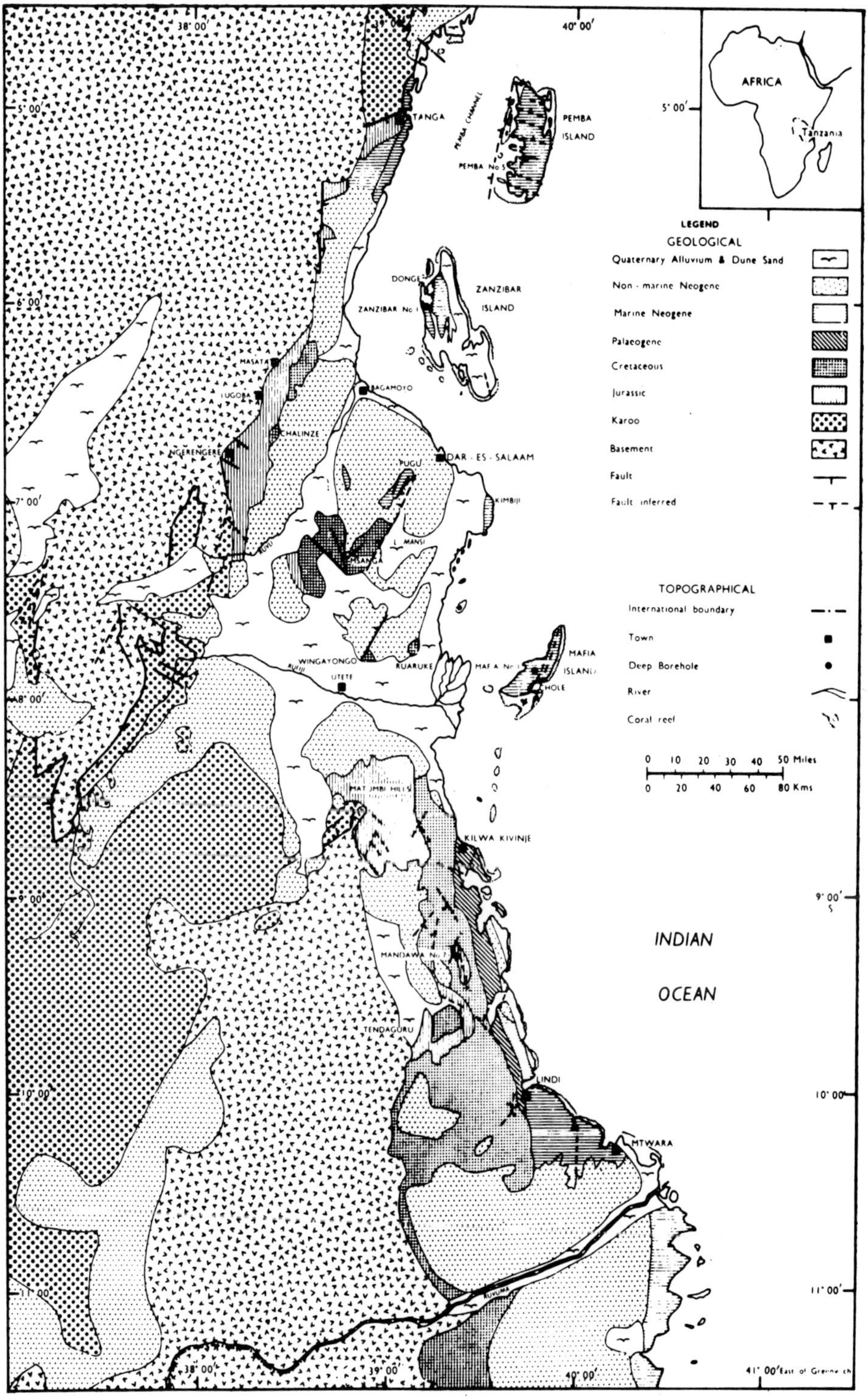

Fig. 9. Map showing geology for coastal Tanzania, based on Kent *et. al.* (1971).

under a service contract covering deep water north of Mafia Island and seaward of AGIP's acreage. A 1000 km deep water reconnaissance seismic survey was shot for Oceanic while AGIP-AMOCO spudded Kisangire-1 close to the western limit of their acreage. Kisangire-1 was drilling below 9200 ft when it was temporarily suspended in February 1976. At the end of 1975, valid petroleum rights in Tanzania covered 97 000 km^2 extending over most of the onshore sedimentary coastal basin, all the shelf and the deep waters north of Mafia Island.

In February 1976 the Tanzania Government granted Oceanic Exploration Company three adjacent blocks covering 40 000 km^2 off Pemba, Zanzibar and Mafia Islands. These new areas, also held under a service contract, are adjacent to the northern half of AGIP-AMOCO's land and shelf block and include areas with water depths ranging from 200 to 3000 m.

In March 1976, AGIP-AMOCO spudded Kisarane-1 in the onshore area west of Dar-es-Salaam and in early 1976 they released their acreage around Songo Songo Island. This was requested by the Tanzanian Government for an appraisal drilling project involving the drilling of an offshore well by TPDC and the Oil and Natural Gas Commission of the Government of India (ONGC). The area relinquished consists of a square block (280 km^2) which included the Songo-Songo-1 well and Songo Island itself. The AGIP service contract granted in 1969 was reduced to 57 000 km^2. Hence Songo-Songo-1 offshore well was completed in 1974 (TD 14 521 ft) in Jurassic rocks after having encountered gas at undisclosed levels. AGIP-AMOCO apparently regard the discovery as non-commercial but Tanzanian sources claim that the find could be commercial.

Subsequently, AGIP-AMOCO released their remaining acreage around Songo-Songo-1 and TPDC and the ONGC became involved in the appraisal of the Songo-Songo offshore find. Drilling, testing and a workover programme (contracted to ONGC) were completed during 1979 and the World Bank recently granted a credit of $30 million for field evaluation. SONATRACH (the Algerian state oil company) contracted in early September 1980 to carry out a drilling programme.

As of September 1980 AGIP (AGIP Africa Optr. 50% and AMOCO Tanzania Petroleum Co. 50%) held 13 950 km^2 in the Songo Songo area south of Dar-es-Salaam. A service contract with TPDC is involved, with all costs borne by the foreign companies and with a 50% state participation in all production.

The total land area of Tanzania is 362 856 square miles; the main sedimentary land area is only 20 077 square miles; the shelf area (0 - 200 m) occupies 8108 square miles and the deep water area (200 - 3000 m) occupies 38 610 square miles (Table 6); herein to some extent lies Tanzania's problem: the smallness of the sedimentary area onshore and the even more limited shelf area currently accessible for exploration and production.

There is fairly good knowledge of well data and a limited amount of seismic data is available publicly in Tanzania, appropriate for a preliminary evaluation. However, as is so often the case, very little detailed seismic data are available publicly and companies which might be interested in exploration are left to speculate about structural and stratigraphic possibilities and prospects.

As in other parts of Eastern and Southern Africa, Karroo sequences were deposited in a series of separate or partially interconnected fault bounded basins formed during the early fragmentation of Pangaea. Major structures were produced mainly by epeirogenic movements and a tensional style of tectonics prevailed until Mid-Jurassic times. Horsts and grabens occur but tilted fault block structures — 'trap door' basins — are extensively developed in Tanzania and Kenya. The Mandawa – Pindiro graben, which extends inland south-south-eastwards from Dar-es-Salaam, and the Songea Karroo trough extending south-westwards from Dar-es-Salaam to Lake Nyasa, together with the trap door fault basins of the Tanga, Pemba and Zanzibar areas, may constitute a privileged junction. Salt seals present in the Mandawa trough, where Karroo coals (?) could be the source of dry gas, make these coastal areas of Tanzania, Northern Mozambique and Southern Kenya areas of potential interest. From Mid-Jurassic times onwards tilting affected what is now coastal Tanzania and Kenya and a typical transgressive – regressive miogeoclinal sequence developed. There was a massive influx of clastics contemporaneous with the uplift of the eastern branch of the East African rift system in Neogene times. Gas plays in the deeply buried Karroo troughs could be of interest in coastal Tanzania.

In the vicinity of Pemba Island, which is thought to be situated over a large horst (perhaps underlain by Mandawa Karroo salt), the Karroo sub-evaporite sequence with sand interbeds lies below 20 000 ft. Further north, or shoreward, it may be present on other highs at more accessible depths and at depths better suited to the generation and preservation of liquid hydrocarbon accumulations in Jurassic, Cretaceous and Cenozoic sands. Limestones may also exist offshore. However, we can but speculate, since no factual data have been published about the distribution of Karroo evaporites, except that these evaporites may have accumulated west of a Basement High which closed certain areas to the invasion of the sea and enabled evaporite basins to develop. No information has been published regarding the volume of matured sediment.

How far eastwards and south-eastwards the Karroo supergroup extended in the subsided continental crust area is unknown but it may well have extended into the Seychelles area and to Madagascar. The plate and break-up history of this portion of the African plate has not been unravelled in sufficient detail to allow definite structural pictures to develop. Figure 9 is an interpretative schematic of the Tanzanian offshore.

Evaluation can only be based on analogy with similar areas when it

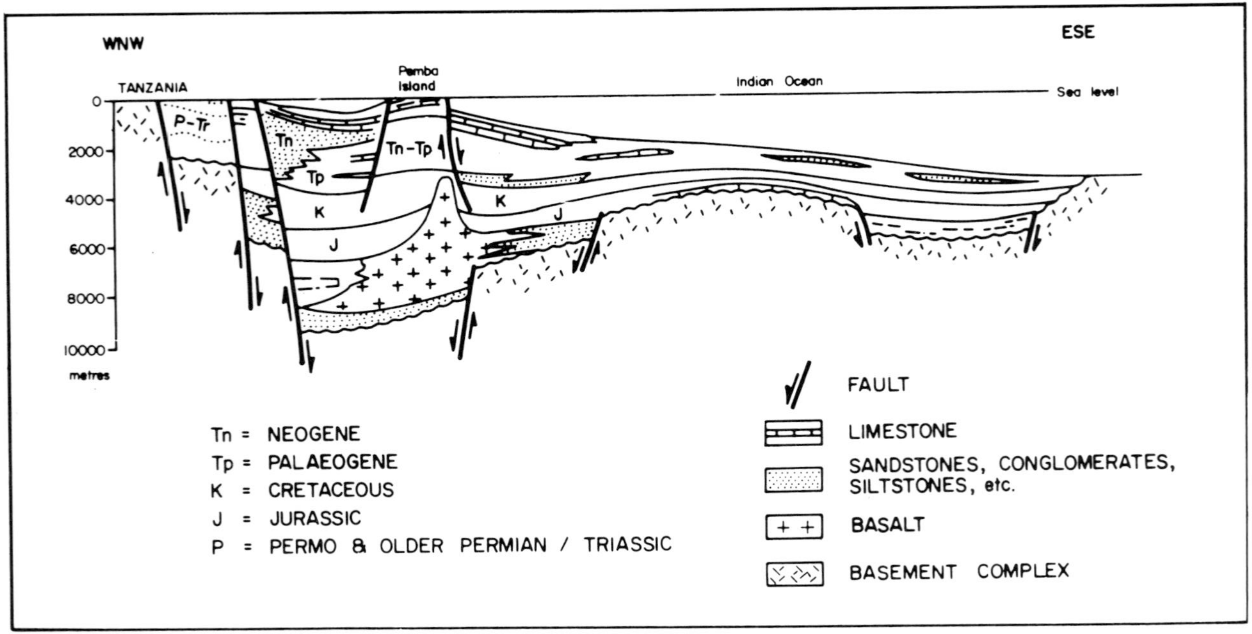

Fig. 10. Schematic geological cross-section through the Tanzanian coastal area.

comes to the 38 000 square miles of deep offshore. Plays may include Tertiary deltaic sands. Growth faults may be present and accumulations would be sourced and sealed by Tertiary shales. Cretaceous shales may also be an additional source. Salt structures could affect Cretaceous and Tertiary reservoirs, providing suitable entrapment conditions.

Deep wells have been drilled on the islands of Mafia, Zanzibar and Pemba but these wells are all dry. Notwithstanding the Songo Songo field evaluation little petroleum activity was reported in Tanzania during 1980, and to date no oil accumulations have been found.

KENYA

General

Petroleum interest in Kenya centres on the southern Kenya coastal belt and the Lamu basin, which, together with the Mogadishu coastal basin and basins farther north in Somalia, form a continuous sedimentary belt.

BP-Shell is the main group to have extensively investigated Kenya's petroleum prospects and that group held licences from 1954 to 1975. The last BP-Shell well was Kipini-1 (TD 12 014 ft) drilled in 1971. The Cities Service group — 33% Cities Service, 33% Marathon Petroleum Kenya and 33% Union Oil Kenya — is the only right-holder (July 1980). The group holds 12 420 km^2 extending onshore and offshore, straddling the coast from Malindi to Patta Island.

The BP-Shell group carried out extensive geological and geophysical work, both onshore and offshore, and drilled light deep wells and numerous shallow stratigraphic holes on the mainland and on the coastal islands.

Kenya's total land area is 225 000 square miles; its sedimentary area is approximately 74 000 square miles, of which shelf area (0 - 200 m) occupies 5500 square miles and deep water areas (200 - 3000 m) occupy about 23 000 square miles. As in Tanzania, the narrowness of the accessible shelf limits prospects. Four areas of interest have been recognized:

(1) Southern Kenya coastal belt and offshore extending southwards into Tanzania;
(2) Lamu embayment, which adjoins the Southern Kenya coastal belt on the south and the Mogadishu basin of Somalia in the north east;
(3) Mesozoic - Karroo basin of north east Kenya, which is part of the Mandera Lugh trough of Somalia;
(4) Kenya deepwater offshore.

Upper Palaeozoic, Mesozoic and Cenozoic sediments occur in these marginal basins which form the Kenyan part of the Eastern African miogeoclinal wedge. At the Kenya coast sediments exceed 32 000 ft on the seaward side of the Lamu embayment.

The sedimentary rocks thin out offshore and at Deep Sea Drilling Project (DSDP) Site 241 some 150 miles from the coast, the sedimentary layer with velocities of 4.7 – 5.3 km/s (characteristic of Kenya Karroo) disappears and the 6600 ft of sediments present consist mainly of Cretaceous rocks (2.5 km/s). At DSDP Site 240, some 500 miles from the Kenya coast, sediments are less than 1600 ft thick.

Prospects of the southern Kenya coastal belt

The narrowness of the shelf in the southern Kenya coastal belt severely limits the prospects. The lack of closed structures is also a problem in this area. Reservoir, seal and source rocks can be identified and there may be pinchout prospects in the Cenozoics. Diapiric salt structures may also exist if the Karroo salt beds extend continuously northwards from Pemba Island and Mandawa in Tanzania (Fig. 2).

Tanzania's Pemba-5 provides clues of what might be expected in the southern Kenya offshore. This well penetrated 10 000 ft of Middle Miocene to Palaeocene sediments but much of the sequence in the lower part consists of siltstones and shales. However, there are promising sands in the Upper Eocene. The southern Kenya coastal belt could have fault controlled structures similar to those postulated for northern offshore Tanzania. Tertiary thicknesses may not be as great as in the Pemba area. Miocene carbonates may form promising prospects but insufficient cover, and adequate seals could be a problem. Cenozoic sands could possibly be viewed as additional targets in these areas and the Cretaceous section, if accessible to the drill, may provide other possibilities by analogy with Tanzania (Fig. 9).

Prospects of the Lamu embayment

BP-Shell wells drilled to date are considered by some geologists to have adequately tested the onshore potential of the Lamu embayment. Seismic data indicate that structural trends recognized on land extend into the offshore. The Lamu basin sequence in Kenya's offshore is still untested but the prospects must be regarded as speculative and high risk. Pre-Tertiary objectives must be at great depth in the coastal area of the Lamu embayment and in the currently accessible offshore because of a thick Tertiary section. The shallower objectives of the Miocene (Fundi Isa) limestones and Tertiary sands are probably the best targets on that part of the shelf. The offshore prospects can only be regarded as highly speculative. The total surveyed part of the area and the subsequently drilled Simba-1 (D and A at 11 824 ft) bottomed in Campanian (Upper Cretaceous) rocks.

Oil seepages have been reported at various times from eastern Kenya but none of them has been checked on the ground. Three wells drilled by BP-Shell encountered gas in small quantities. In Dodori-1 small amounts of dry gas were produced from the Middle Eocene and some wet gas and salt water from the Lower Eocene and Palaeocene. Palaeocene sands are reported as bitumen-bearing in this well.

Sandstone and carbonate rocks (Upper Cretaceous to Miocene) with reservoir characteristics have been proven in all BP-Shell wells drilled in the Lamu embayment but porosity and permeability are said to deteriorate with depth. Lower Eocene to Miocene reservoirs proven by drilling generally have good characteristics. Palaeocene reservoir characteristics, as seen in Dolori-1, are generally fair, whereas Mesozoic reservoirs, with the exception of the gas-bearing Upper Cretaceous sands found in Kipini-1, have less well developed characteristics. Seals are described as generally adequate, although in the Miocene limestone sequence near the coast shales are only thinly developed. Shale/clay development in the Palaeocene to Oligocene section is generally reported as adequate for entrapment.

Walters and Linton (1973) did not comment on source materials and no maturation data have been published. Various lignitic occurrences may be considered possible gas sources and the Upper Jurassic and Upper Cretaceous shales could be possible source rocks for oil.

Small amounts of dry gas were produced from the Middle Eocene in Dodori-1 while some wet gas and salt water were produced from the Lower Eocene and Palaeocene. Bitumen indications were found in the Palaeocene limestones. Drill stem tests yielded small quantities of dry gas from Middle Eocene sandstones in Pandangua-1. In Pate-1, gas under pressure was encountered in three thin sand layers in the Lower Eocene. The gas was initially wet and had traces of condensate but it became leaner after flowing for some time. Small amounts of dry gas were found in Kipini-1 in the Middle Eocene. These shows are associated with coals and/or lignitic beds. Senonian sands totalling 36 ft in thickness yielded wet gas in Kipini-1. Specks of oil were also isolated from the drilling mud and patchy fluorescence was noted in cores.

BP-Shell wells drilled to date are considered to have adequately tested the onshore Cretaceous-Cenozoic potential of the Lamu embayment but seismic data indicate a prolongation of the land structural trends under more favourable conditions. Offshore prospects are discussed below. If the Karroo fault trough basins of Tanzania and the southern coastal belt of Kenya extend northwards to join the Karroo Mandera Lugh trough of Somalia beneath the Lower Cretaceous and Middle and Upper Jurassic of the Lamu embayment, as has been postulated, there might be Karroo prospects accessible at economic depths in the Lamu embayment. A schematic and somewhat hypothetical section across part of the Lamu embayment is presented in Fig. 11.

BP-Shell has not published any maps showing the depth to the basement complex for the Lamu embayment, and the structure contours plotted on the top of the Palaeozoic and shown on the Association of African Geological Surveys 1969 Tectonic Map of Africa (scale 1 : 5 000 000) are misleading because of the way in which the data

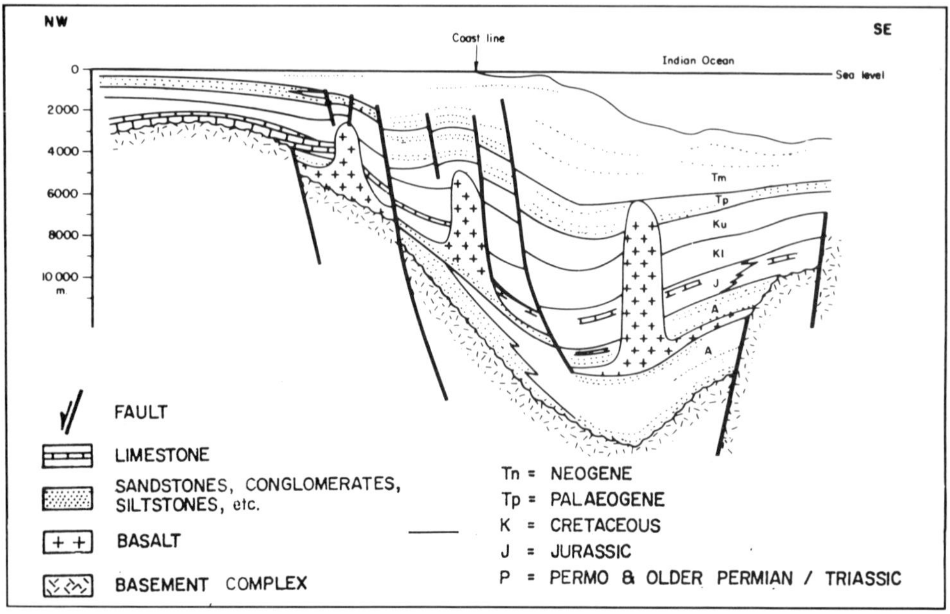

Fig. 11. Schematic geological cross-section across the Kenyan coastal area.

were collected. Therefore they cannot be considered proof of the facts concerning Karroo sediments in the Lamu embayment.

Structural indications are not readily apparent in the Lamu embayment as the surface is a gently undulating to an almost flat featureless sand-covered plain. In the southern part of the embayment the dips are gentle (around 5°) and to seaward, dip reversals are rare. Faults trending north east and south west are common and local pre-Campanian highs trending in these directions are known. Other trends also occur below the Campanian. The BP-Shell aeromagnetic servey confirmed the boundaries of the embayment and confirmed that a basement ridge runs east–west south of Wajir, separating the Lamu embayment to the south from the Mandera basin of Somalia to the north. A gravity ridge was defined east of the Tana River running roughly north–south from Bura-1 through Walu-1, Pandangua-1 and Kipini-1 (Fig. 12).

Two separate gravity highs were defined north of the Walu-1 high more or less along the Tana River and centred on Garissa-1. North east of the Walu-1 gravity high another high was defined around Wal Merer-1. BP-Shell has reported that several of the gravity anomalies were investigated by seismic reflection surveys and that reflection surveys were also carried out elsewhere irrespective of gravity highs. No details have been published, as in Tanzania (Kent *et al.*, 1973), and the Kenya Geological Survey holds this information 'tight'.

As is common in Eastern Africa, there is no compressional folding. Some drape structures may be developed over salt diapirs, over faulted

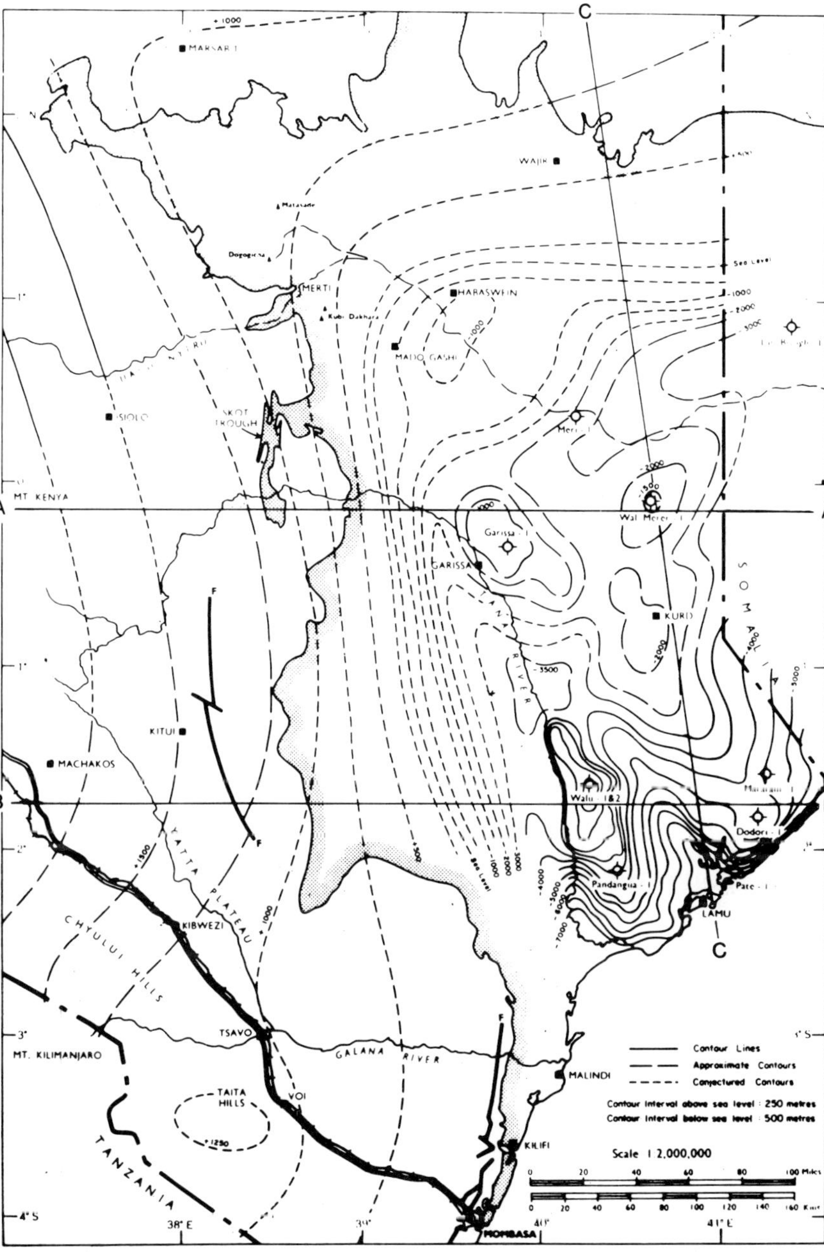

Fig. 12. Map showing 'structure' contours on the top of the 'End Tertiary Erosion Surface' or on the top of the Mesozoic sediments. Based on Walters and Linton (1973).

older strata and basement complex blocks.

Offshore, if salt is present, then faulted structures of the Pemba Island type related to salt diapirs may exist but, as we have already said, evidence of salt and evaporites is speculative further north than the Southern Tanzanian Mainland, although it is said to be supported by existing seismic data (Fig. 11).

Four distinct periods of faulting have been recognized in Kenya and their intensity has varied with time:

(1) Karroo faulting (Carboniferous to Pre-Mid-Jurassic);
(2) End Cretaceous faulting;
(3) Oligocene faulting;
(4) Pliocene faulting.

Karroo faulting affects the Southern coastal belt of Kenya and adjacent parts of Tanzania. Whether Karroo fault troughs occur beneath the Lamu embayment is uncertain. Karroo faults do occur, though, in the Mandera Lugh trough of Somalia. Two major structural trends recognized onshore extend into the adjacent offshore. These are:

(1) the Walu - Pandangua - Kipini anticlinal trend;
(2) the Patta - Dodori - Mararani anticlinal trend.

Prospects of the northern Kenya basin - Mandera Lugh trough

Very little is known about the stratigraphy and structure and consequently the petroleum endowment of north east Kenya. An ill-defined Mesozoic basin forming part of the Mandera Lugh trough of Somalia and Ethiopia may extend over northern Kenya and join with the Lamu embayment. The Mandera Lugh Trough is an extensive Karroo trough system formed during Late Carboniferous to Early Jurassic times and best defined in Somalia and Ethiopia. Its continuation into Kenya is uncertain. Continental clastics in substantial amounts could exist sealed in by Late Jurassic - Early Cretaceous and younger rocks if the Mandera Lugh trend continues beneath the Lamu embayment to the Mombasa - Malindi Karroo rocks which occur near the coast.

Prospects of the Kenya deepwater offshore

Very few data are available for assessment (either qualitative or quantitative) but, by analogy with the onshore, and in a regional context, it may be said that Palaeocene - Miocene multiple sands and limestone plays could be expected, sealed and sourced by shales of the same ages.

Some 3500 line km of seismic surveys were shot by Total in 1973. Subsequently Total drilled Simba-1 (TD 11 824 ft) in Campanian.

Only one well has been drilled in Kenya's offshore, and no data have been released. Much of the older Karroo formation and the Jurassic and Cretaceous section are likely to be inaccessible even in the shallow offshore.

Certainly for Middle and Late Jurassic, Cretaceous and Cenozoic sediments, there is a normal decrease in grain size which drastically reduces the prospective interest. Turbidity and seaward contourite sands could be expected but their prospects are highly speculative. The Karroo salt features shown in Fig. 11 are also purely speculative. Without doubt the Cenozoic section thickens dramatically offshore as is clearly demonstrated by Pemba-5. Pemba-5 penetrated about 10 000 feet of Middle Miocene to Palaeocene sediments, and much of the lower part of that section consists of siltstones and shales. Promising sands may exist in the upper part of the Eocene.

The offshore of the southern coastal belt of Kenya and that of the Lamu embayment could have similar fault controlled features and a similar stratigraphy to offshore and coastal Tanzania. Miocene carbonates could act as promising reservoirs but the questions of seals and sufficient cover may be critical for accumulation. Sandstones within the Tertiary can be an additional target, but the limitations of seals and lack of cover could once again be critical. Cretaceous and older rocks may well lie too deep for economic drilling in the offshore area of the southern coastal belt of Kenya.

In some areas, the Continental shelf of the Lamu embayment widens to almost 40 miles, but owing to a large and thick onshore section in the Lugh embayment it follows that thick Tertiary strata can be expected offshore. Data collected by a scientific survey ship confirm this to some extent. Consequently, pre-Tertiary objectives may be buried at too great a depth to be regarded as economic drilling projects. Furthermore, the depth may have destroyed the porosity and permeability or temperatures may be too high for oil or gas to have been preserved. Shallower objectives such as the Miocene Fundi Isa limestone and Tertiary sands should be the best targets on the offshore Lamu shelf.

A number of closed structures are known to exist in the offshore area of the Lamu embayment on the flanks of both the Walu–Pandangua-Kipini trend and the Mararani-Dodori-Patta trend. A well drilled on the Kipini structure in the shallow offshore may encounter prospects in Middle and Upper Eocene porous limestones sealed in by Oligocene shales, etc. and in Lower Eocene and Palaeocene deltaic sands where good reservoir conditions can be expected from offshore experience. Good marine source rocks could exist down a dip and towards the south east. Marine shales are almost certainly well developed in the underlying Senonian (hydrocarbon occurrences of Kipini-1) and hydrocarbons could have migrated into Palaeocene and Lower Eocene reservoirs.

SOMALIA

Again it was 'seep' oil that drew the attention of early prospectors to Somalia and a report on the Dagah Shabell oil seep was made as early as 1912 by government officials.

From 1944 to 1951 the Somaliland Exploration Company consisting of Anglo-Egyptian Oilfields, Shell Overseas Exploration Company and D'Arcy Exploration Company (BD) held exploration licences over what was then the British Somaliland Protectorate. No production was established from these investigations; a few seeps were found. After independence, in 1960, there was renewed interest in Somalia and up to 1975 48 wells had been drilled but without success. Between 1975 and 1979, three more wells were drilled, with no more success.

The total area of Somalia is 270 000 square miles and main sedimentary areas cover 154 000 square miles. The shelf from 0 to 200 m covers about 23 000 square miles and deep water (200–3000 m) covers around 115 000 square miles (Table 6). Most of the prospective offshore area lies on the Indian Ocean side of Somalia because the shelf and slope of the Gulf of Aden protomiogeocline are very narrow and steep. Sediments comprising the Indian Ocean miogeocline exceed 30 000 ft in thickness. There are three main areas of interest to prospectors for oil and gas in Somalia:

(1) Northern coastal basins (onshore and offshore) bordering the Gulf of Aden and involving in large-scale vertical movements consequent upon lithospheric plate thinning (leptogeoclinal phase); plate separation (late leptogeoclinal phase); seafloor spreading (protomiogeoclinal phase) and the development of the Gulf of Aden (Sheba Ridge) ridge-transform system.
(2) Somalia Indian Ocean coastal basins developed extensively offshore and onshore. These include:
  (a) Socotra shelf area (South Yemen) and Cape Gardafui shelf (Somalia);
  (b) Ras Hafun basin;
  (c) Migiurtina (Cotton) basin with depths to basement complex estimated at greater than 15 000 ft;
  (d) Ogaden basin or Somalia embayment (mainly Mesozoic) extending into Ethiopia.
(3) Mandera Lugh trough developed north west of the Bur Acaba basement high which limits the Mogadishu basin on the west. This trough, which formed mainly in pre-Late Jurassic (Karroo) times, extends northwards into Ethiopia (Whiteman, 1975) and which may extend southwards beneath the Lamu basin of Kenya and into the coastal Karroo fault zone and troughs of Southern Kenya and Tanzania.

Parts of Somalia fulfil all the requirements of a potential hydrocarbon province, as hydrocarbons have been generated in Jurassic, Cretaceous and Tertiary rocks and in the Somali section of the Eastern Africa miogeocline the sedimentary column exceeds 30 000 ft. Many instances of porosity and permeability are known of both the carbonate and clastic type. However, oil and gas have not been found to date in Somalia.

The areas with most prospective interest in onshore Somalia are thought to be the Somalia embayment or Ogaden basin characterized by Mesozoic shelf and reefal sediments, and the area around the Nogal uplift. Other areas of interest are the Ras Hafun trough and the Indian Ocean offshore, which is largely unexplored. Most of the surface anticlinal features known onshore are said to have been drill tested in the last 20 years without commercial accumulations of oil being found. However, this excludes the Mandera Lugh basin, where drilling activity has been lower than in other basins and where interest has only recently started to build up again.

Prospectors recommend that a stratigraphic approach might well be more profitable than the structural approach (Barnes, 1976). Stratigraphic traps may occur in clastic sedimentary rocks where facies variations are widespread and common. Unexplored structural prospects exist offshore and in parts of the onshore and, as we have said, in the Mandera Lugh trough system.

Marine geophysical data, mainly from an academic survey ship, indicate that offshore sediments thin eastwards and form part of the great Eastern Africa miogeoclinal wedge. The greatest volume of offshore sediments occurs along the Somali Indian coast, where Mesozoic – Cenozoic sediments exceed 30 000 feet.

Survey ship geophysical data indicate that offshore in the Gulf of Aden sediments thin out more rapidly than on the Indian Ocean side, as they form part of a mainly Neogene protomiogeoclinal wedge which developed after separation as new oceanic crust was generated and formed the Sheba Ridge transform system extending from the Gulf of Tadjura in the west to the Owen fracture zone in the east. As a consequence of the narrowness of the shelf, the steepness of the slope and the thinness and possibly the immature nature (mainly Neogene) of the sediments, the prospects for the onshore-offshore on the south side of the Gulf of Aden are considered to be poor.

Prospects exist, however, in the eastern part of the Gulf of Aden in the area extending from Cape Gardafui to Abd al Kuri Island, Brothers Islands, and Socotra (South Yemen). This area really forms part of the Eastern African miogeoclinal wedge.

Prospects of the Cape Gardafui (Somalia) – Socotra shelf

Esso/Elf Aquitaine held interests offshore and onshore in this area of Somalia where large thicknesses of Mesozoic-Cenozoic rocks are known. The estimated depth to basement is around 16 000 ft. Nearby in the Socotra area 1600 km of marine seismic survey were shot and 6400 line km of aeromagnetic data recorded in December 1975. Well

Gardafui-1 (Elf) was spudded on 22 September 1975 on the Gardafui shelf. No finds have been reported to date.

Prospects of the Ras Hafun basin

The area is covered by Neogene sediments and evidence has been reported which indicates that a major trough infilled with Miocene sediments exists in the area. This rift trough is thought to trend east–west and may well be part of the Scusciuban rift zone situated some 75 - 150 miles to the west. Distinct rift type faults are known in this region. The Hafun graben is marked by a gravity low. Eocene and Oligocene outcrops mark synclinal areas in the coastal belt and a large anticlinal structure has been reported south of Ras Hafun.

Wells drilled offshore are said to have proven very substantial thickness of attractive reservoir rocks and substantial seals in formations which could act also as source rocks.

Hordio-1 drilled onshore by Min. Somali (AGIP) is said to have encountered three porous zones in the Middle and Upper Cretaceous and a carbonate unit is reported to have tested salt water and some gas.

Prospects of the Migiurtina or Cotton basin

Offshore, the Ras Hafun basin merges southwestwards and eastwards with the Migiurtina basin. Seven Coronada and three AGIP wells drilled in this basin all proved dry. Very little stratigraphic data have been released, which is an impediment to our knowledge of the area.

Well Sagaleh-1 drilled on the southern margin of this basin by Min. Somali reached basement at 10 722 ft. Three porous carbonate zones were encountered in the Middle and Upper Cretaceous. The upper zone is said to be around 800 ft thick, the middle zone around 230 ft, and the lower zone around 320 ft. In addition, a 950 ft thickness of very coarse and porous Liassic sandstone was drilled, resting on the basement complex. Oil stains were found in a porous Jurassic dolomitic limestone in Sagaleh-1 and in the Auradu limestones.

Wells in Hordio-1 and Cotton-1 are both reported to have had gas shows in the Upper Cretaceous. The gas shows in Cotton-1 span a 300 ft interval in Cretaceous limestones. Beneath an unconformity at the base of the Cretaceous, there are Callovian - Oxfordian sediments and Liassic dolomites, limestones and anhydrites. Oil stains were also found in the Auradu limestone in Cotton-1. Some of the Jurassic sequence is thought to have been deposited in a restricted environment and these rocks may have acted as source beds according to Elf.

Prospects in the Ras Hafun basin are considered to have some attraction but are classed as speculative, especially offshore. The shelf is narrow and deeper water prospects are uncertain. There may be prospects in Upper Mesozoic and Tertiary sands associated with fault controlled features and there may also be salt induced structures, bearing in mind the presence of anhydrite mentioned above.

Prospects of the Ogaden basin

This basin was explored extensively by Sinclair and Tenneco mainly in adjacent Ethiopia. Sinclair had little success but Tenneco recently reported a gas condensate discovery in Calub-1 (TD 12 139 ft) in this basin. It is reported that Calub-1 flowed around 3.5 MMCFD and also had oil shows in the Jurassic. This find focuses interest on the Somali section of the basin. Sinclair's Gira-1 had a small gas show in the Jurassic and the Auradu limestone was found to be oil stained.

Prospects of the Mogadishu coastal basin

The Bur basement high complex located about 60 miles west of Mogadishu separates southern Somalia into two structural units — the Mandera Lugh basin or trough on the western side of the high, and the Mogadishu or southern Somalia coastal basin on the eastern side (Fig. 13).

Geophysical and stratigraphic evidence suggests that the Bur Acaba region was uplifted by regional epeirogenic movements near the end of the Jurassic period. During the Early Cretaceous deposition continued in the coastal basin and active subsidence was controlled by north east - south west trending faults bounding the Bur Acaba high.

Surface information about the stratigraphy of the Somali coastal basins is screened by flat lying Quaternary sediments. A thick Jurassic sequence has been preserved on the downthrow side of a major fault with a more than 14 000 ft throw situated on the eastern side of the Bur Acaba high. This fault zone may extend more or less continuously north-eastwards along the Somali coast.

Subsurface data indicate that there are three major stratigraphic units of prospective interest in the Mogadishu coastal basin. From top to bottom, they are:

(1) Miocene Somali formation, consisting mainly of carbonates with occasional clastics resting with angular unconformity on older beds. Some sandstones occur and thickness ranges from 3000 to 4000 ft.
(2) Early Tertiary Lach Dera formation, consisting of terrigenous sediments, mainly quartzose sandstones with shales and mudstones and lignites in the upper part. Thickness varies from 7000 ft in Lach Bissign-1 to over 9000 ft at Oddo Alimo-1. A marine intercalation occurs in this well.
(3) Upper Jurassic - Lower Cretaceous Brava formation, composed almost entirely of shale. Around 9500 ft of this sequence was drilled in Sinclair's Brava-1. The basal section becomes sandy and there are around 2000 ft of quartzitic sandstones with shales.

Deposition in the Mogadishu coastal basin, as in the Lamu basin of Kenya which it adjoins, was controlled by separate fault blocks. At the coast near Mogadishu, the Tertiary is reported to be as much as 2500 ft thick.

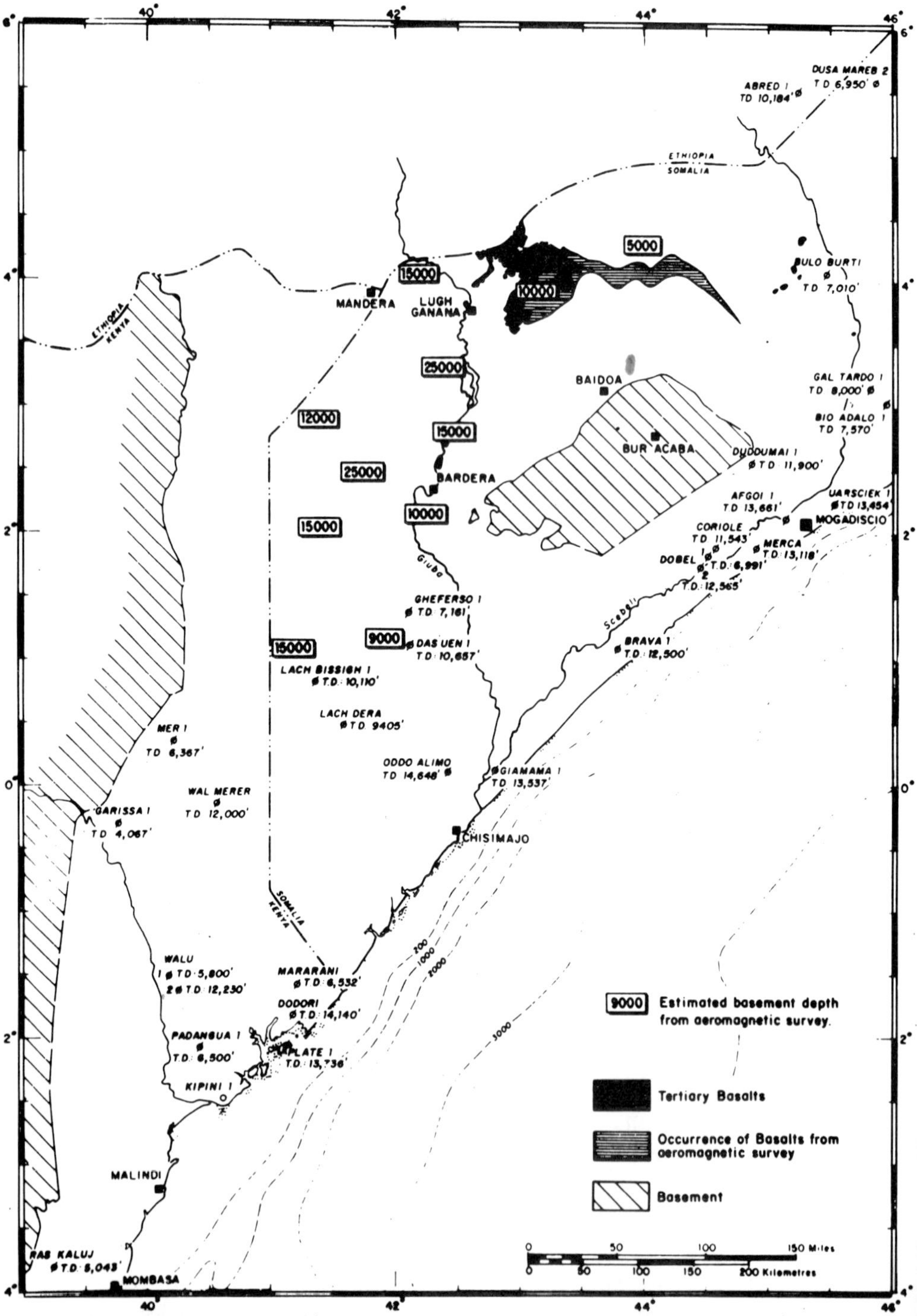

Fig. 13. Map showing depth to basement (estimated largely on aeromagnetic data) Bardera and Mandera Lugh areas, Somalia and Ethiopia. Based on Beltrandi and Pyre (1973).

Depth to the continental basement in the Southern coastal basin may exceed 48 000 to 50 000 ft in places. The Cenozoic might be as much as 25 000 ft thick and strongly diapiric salt structures might be present affecting both Mesozoic and Cenozoic strata. There is no published information about whether or not the Juba and Shebele (Vebi Giuba and Vebi Scebelli) rivers, which are ancient primary drainage lines and which had established delta complexes in the Palaeogene, developed offshore delta complexes in the Palaeogene–Neogene.

Around 20 wells have been drilled north east and south west of Mogadishu in the onshore part of the basin. All were dry, although Afgoi-1 was reported as a non-commercial gas discovery. One horizon is reported to have produced 110 bbl of distillate and another 5 – 10 MMCFD of gas. Three other wells drilled on the Merca and Cariolei structures are reported to have had oil and gas shows.

Eocene Sandstones in Merca-1 were oil stained and contained gas under good pressure. In a drill stem test of the depth interval 8869 to 8875 ft salt water flowed at the rate of 15 gallons per minute accompanied by 25 – 30 MCF of gas. Bottom hole pressure was 5500 psi. Salt water again flowed at 7858 to 7870 ft with a pressure of 4900 psi on a 3/8 inch choke.

Offshore Upper Mesozoic and Cenozoic sands associated with fault controlled structures (possibly related to deep diapirism) may be prospective.

Summary of hydrocarbon prospects, Mandera Lugh trough

The Mandera Lugh trough is situated on the west of the Bur Acaba basement high. It is an extensive Karroo trough of Late Carboniferous to Early Jurassic age (Fig. 14). Pre-Jurassic formations do not crop out in western Somalia but regional stratigraphic evidence from Kenya and geophysical evidence from Ethiopia and Somalia (Table 6) indicate that substantial thicknesses of predominantly continental clastic pre-Jurassic rocks exist in the Mandera Lugh trough of Ethiopia and Somalia. The total thickness of Mesozoic rocks measured on the south east flank of the Somalian part of the basin and along the western flank in Kenya averages 8000 and 10 000 ft respectively. However, aeromagnetic data indicate a thickness of 25 000 ft along the basin axis.

Within the Mandera Lugh basin, the Jurassic totals 8000 ft and appears to have been deposited in a trough initiated in pre-Jurassic (Karroo) times and forming part of an extensive series of Karroo troughs which may have extended from the Red Sea to Tanzania.

In early Jurassic times, the trough was invaded by the sea and the neritic carbonates of the Didinitu Beds were laid down. Progressive subsidence followed in Mid-Jurassic times and a thick sequence of limestones with minor shales (Baidoa formation and equivalents) were deposited. Towards the end of Mid-Jurassic times and in Late Jurassic times, more detrital neritic sediments (Miritini and Rabi shales of

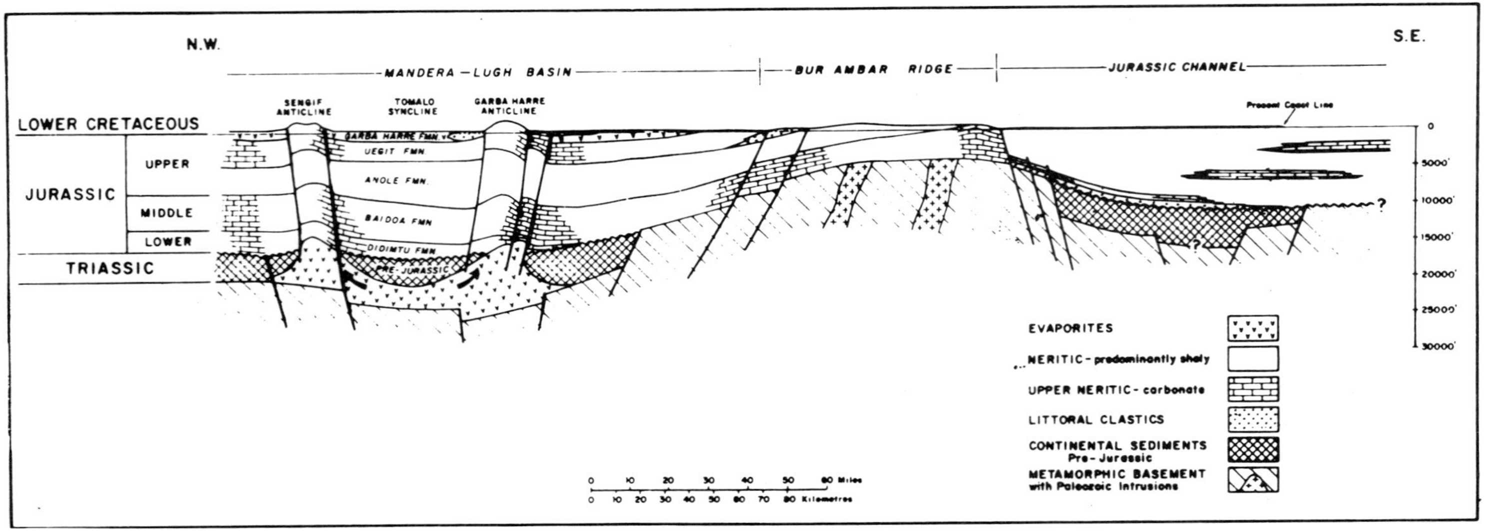

Fig. 14. Hypothetical structural and depositional section constructed to end of Early Cretaceous time for the Mandera Lugh Trough, Bur Amba Ridge and Somalia Coastal Channel (Jurassic Channel). Note the postulated movement of basal pre-Jurassic evaporites in the Mandera Lugh Trough in the Somalia Coastal Area. Based on Beltrandi and Pyre (1973).

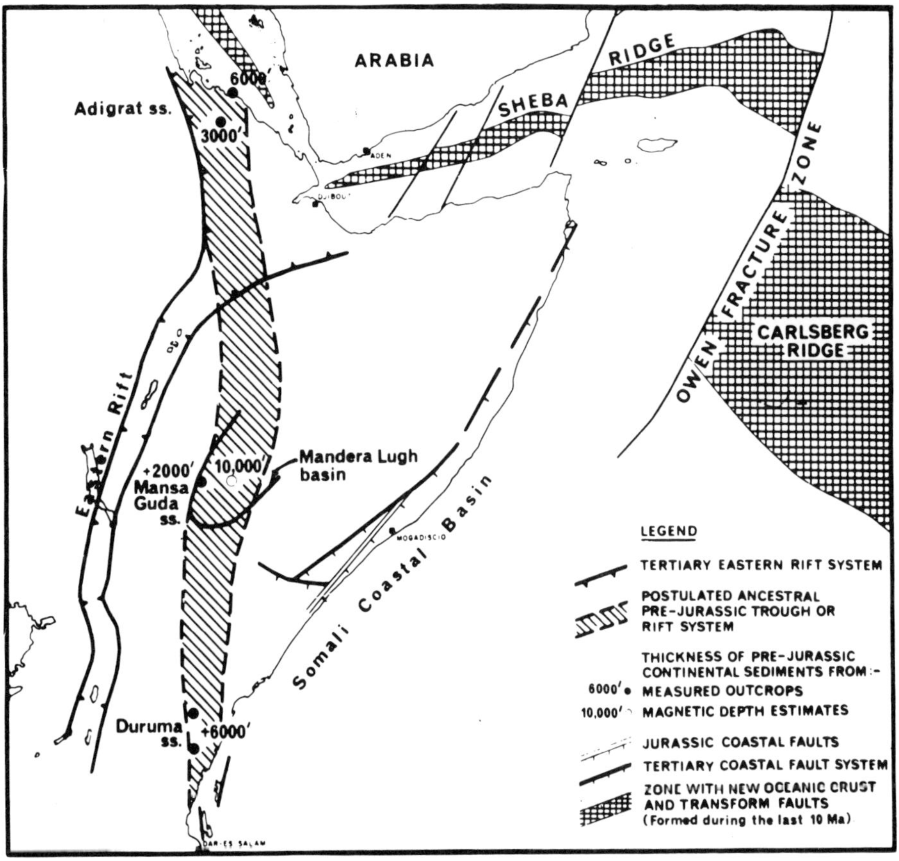

Fig. 15. Map showing the relationship of the Mandera Lugh Karroo Trough (pre-Jurassic), the Cenozoic Eastern African Rift System and the Eastern African Miogeocline. Based on Beltrandi and Pyre (1973).

coastal Kenya, the Rukeso and Ramu shales of north east Kenya, and the Anole formation of Western Somalia) were deposited.

Widespread regression was initiated as a result of epeirogenic uplift of the trough system, especially along the western margin of the trough and the quartz sandstone of the Uegit formation was laid down.

Towards the end of the Jurassic period, the Bur Acaba basement high acted as a physical barrier between the Mandera Lugh trough and the Somali coastal basins. In Late Jurassic and Early Cretaceous times individual evaporite basins formed in the Mandera Lugh trough and continental sandstones were laid down during a period of structural uplift.

Burmah Oil's Hol-1 tested gas in the Jurassic. The lower carbonate unit (Galoda Member) was a prime exploration target and the basal sandstone Deleb Member and Uegit Member (upper carbonate unit) are reported to have been secondary targets. All these units may have considerable reservoir potential. Seals in the form of thick shales (Anole formation) exist in the Hol area and they could also be source beds. Four structures appear to have been defined in the area: (1) tossile structure; (2) Hol Anticline; (3) El God dome; (4) Cursi structure.

Tenneco tested significant but apparently non-commercial oil and gas on the El Kuran structure. El Kuran-1 was drilled to TD 10 462 and El Kuran-2 to TD 6610 ft.

In assessing the hydrocarbon potential of the Mandera Lugh basin (10 038 square miles or 26 000 km^2), we have a fair knowledge of the stratigraphy and structure from the published literature and various farm-in proposals. We know there is a well-developed Karroo rift trough, the major part of which is in Kenya and Ethiopia, but the eastern side appears to lie in Somalia. Burmah Oil's Hol-1 probably was drilled in the trough sediments.

Some 6000 to 10 000 ft of shallow marine Cretaceous and Jurassic sediments overlap Triassic evaporites which form faulted salt pillows of diapirs at depth and oil and gas may occur in oolitic limestones trapped on fault structures. Oil could be expected in basal Jurassic sands sealed in by shales and trapped in fault blocks. Gas, condensate and oil might be expected from finds made to date, but now the maturation data have been published, the remoteness of the area detracts from its potential interest. The area is thought to have low to moderate potential but must be rated as high risk. Nevertheless, there may be fair rewards.

SELECTED REFERENCES

American Association of Petroleum Geologists (1954-1980). Foreign Developments. *Am. Assoc. Petrol. Geol. Bull.*

Anon (1980). Reserves. *Oil Gas J.* 29 December 1980.

Barnes, S.O. (1976). Geology and Oil Prospects of Somalia, East Africa. *Am. Assoc. Petrol. Geol. Bull.* **60** (3), 389-413.

Beltrandi, M.A. and Pyre, A. (1973). Geological evolution of southwest Somalia. In *Sedimentary Basins of the African Coasts* (Ed. G. Blant), Assoc. African Geol. Surv., Paris.

Coury, A.B. and Henricks, T.B. (1978). Europe, West Asia and Africa. Map of prospective hydrocarbon provinces of the world. *Misc. Field Studies.* Dept. of Interior, USGS.

Flores, G. (1970). Suggested origin of the Mozambique Channel. *Trans. Geol. Soc. S. Afr.* LXXII.

Grossling, B. (1976). Window on oil. *A Survey of World Petroleum Resources.* Financial Times, London.

Kent, P.E. (1965). An evaporite basin in Southern Tanzania. In *Salt Basins Around Africa.*

Kent, P.E., Hunt, J.A. and Johnstone, D.W. (1971). Geology and geophysics of coastal Tanzania. Institute of Geological Sciences, Gt. Britain. Geophys. Paper 6.

Kent, P.E. and Perry, J.T. (1973). The development of the Indian Ocean Margin in

Tanzania. In *Sedimentary Basins of the African Coasts* (Ed. G. Blant), Assoc. African Geol. Surv., Paris.

OMNIS — Petroconsultants (1980). Bidding documents for promoting exploration in the Majunga and Movandava Basins, Madagascar. Office Militaire National pour les Industries Strategiques.

Petroconsultants, S.A. (1980). Kenya Coastal Area Map 1 : 2 000 000. Foreign Scouting Service. Petroconsultants S.A. Geneva. July 1980.

Petroconsultants, S.A. (1980). Tanzania Coastal Area Map 1 : 2 000 000. Foreign Scouting Service. Petroconsultants S.A. Geneva. September 1980.

Rigassy, D.A. and Dixon, G.E. (1972). Cretaceous of Cape Province Republic of South Africa. *Proc. Conf. African Geology*. Ibadan Univ. Nigeria, 1970 (Eds A.J. Whiteman and T.F. Dessauvagie).

UNESCO-AAPG (1980). Tectonic Map of Africa. Scale 1 : 5 000 000.

Walters, R. and Linton, R.E. (1973). The sedimentary rocks of coastal Kenya. In *Sedimentary Basins of the African Coasts* (Ed. G. Blant), Assoc. African Geol. Surv., Paris.

Whiteman, A.J. (1971). *Petroleum Geology of South Africa*. Gulf Oil Company Internal Report.

Whiteman, A.J. (1975). Geology and hydrocarbon prospects of the Red Sea and Gulf of Suez. Exploration Consultants Ltd., UK, 1-682. 363 figs.

Whiteman, A.J. (1977). Geology and Hydrocarbon Prospects of Eastern Africa, Cape of Good Hope to the Horn, Somalia. Open Report. Exploration Consultants Ltd., UK. Vols I-IV, 1200 pp., 650 figs.

PART II

Petroleum Exploration Technologies

5

PETROLEUM EXPLORATION METHODS, TECHNIQUES AND COSTS

P. Kassler

Head of Exploration/Production Economics, Shell, The Hague, The Netherlands

INTRODUCTION

It gives me great pleasure to have the opportunity to address this distinguished international gathering of experts on a subject which happens to be of particular concern to all of us — that of exploration for oil and gas resources in areas in which up to now they have not been discovered, or perhaps only in commercially insignificant quantities. I should like to start by saying that although this topic is often in the political arena, the central problem is a technological one, i.e. the occurrence of commercial quantities of oil and gas is determined solely by complex geological factors, and cannot be influenced by government policy.

The work of exploration consists therefore of selecting and applying the most appropriate scientific and technical methods to find petroleum. However, in spite of using the best available methods, exploration often fails to discover commercial oil or gas, and, whether successful or not, exploration is very costly in terms of cash and skilled manpower.

This is the problem underlying the task or professional activity known as 'exploration management', which consists of optimally allocating scarce resources (people, cash and hardware) in order to determine the existence, geographical location and size of petroleum reserves.

My comments today are made from the particular viewpoint of an international oil and gas company, and I make no apology for that fact. It is evident, however, that efficient exploration management is a requirement for any organization charged with petroleum exploration in a particular country, whether it is primarily responsible to its government or to private shareholders. In either case the expected benefits from the proposed exploration activity, such as the value of petroleum resources to be discovered, import savings, etc. should exceed the expected resource cost of the exploration itself if the exploration is to be justified. The real problem, of course, is that neither the benefit nor the cost can accurately be determined in advance of making the exploration investment, and that even with great technical competence one can do no more than make some reasonable judgements about the probabilities of success and the probable costs of exploration. This is incidentally one of the main difficulties in negotiating exploration contracts with governments, in which one is often obliged to back these judgements with hard cash in the virtual absence of any real geological information.

A further dimension is added to the petroleum exploration problem when one considers different categories of countries or geological provinces. It is well known that petroleum is distributed in a very uneven manner, some countries being extremely rich (e.g. Saudi Arabia) while others have very little (e.g. Sweden). In many of the developed countries, particularly in Western Europe and North America, economic conditions have for many years encouraged petroleum exploration, and it is fair to say that their potential is reasonably well known. In many of the developing countries this strong economic motivation has been slower to appear. Consequently there are time lags in the acceleration of exploration, and the degree to which particular countries' reserve potential has been fully evaluated is quite variable. Some countries have certainly not yet been fully explored, for either geological, logistic or internal political reasons, i.e. oil and gas may be absent for geological reasons; petroleum reserves may be present, but technically difficult or expensive to discover or far from potential markets; surface conditions may have impeded exploration; potential explorers may not have been

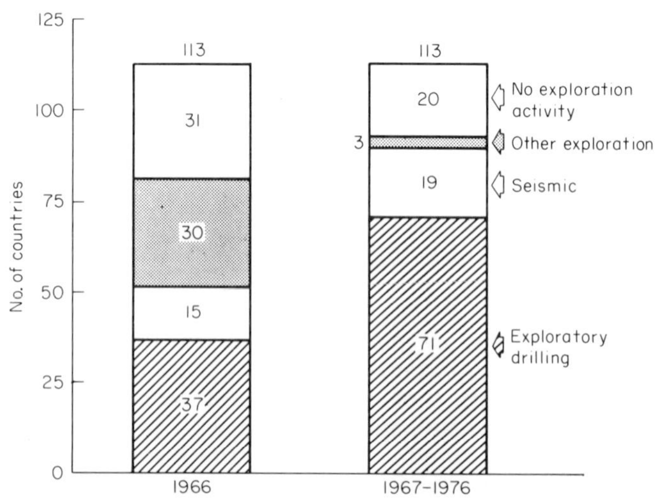

Fig. 1. Extent of exploration activity in non-OPEC developing countries.

able to gain access to do exploration because of the absence of suitable legislation, etc.

In spite of all this, it is striking how many developing countries have by now been at least partially explored. This is shown in Fig. 1, which particularly indicates the enormous expansion of exploration work that has occurred outside the major producing countries in the past few years. This chart indicates that only 20 countries had not been explored by 1976. If those countries in which the unfavourable geology practically precludes the presence of petroleum are excluded, the number of unexplored countries which still have genuine prospects today is probably very small.

In these comments I have deliberately based myself on the view that 'geology works' and that by the judicious use of scientific methods and techniques the chance of finding petroleum can be greatly increased, compared with, for example, random drilling. This further implies that the richness of some oil provinces and the leanness of others is due to geological factors which can be studied, and to some extent predicted,

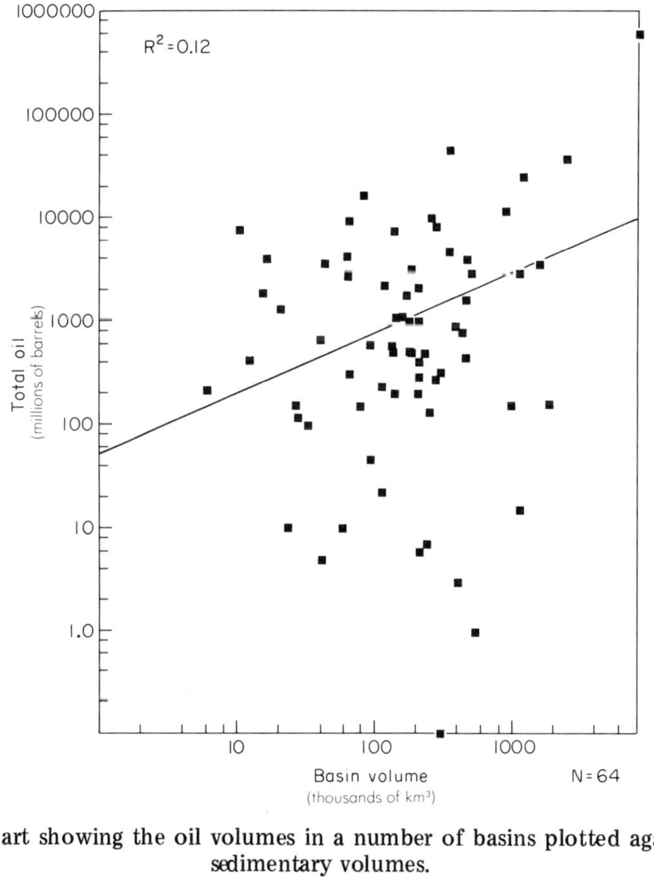

Fig. 2. Chart showing the oil volumes in a number of basins plotted against their sedimentary volumes.

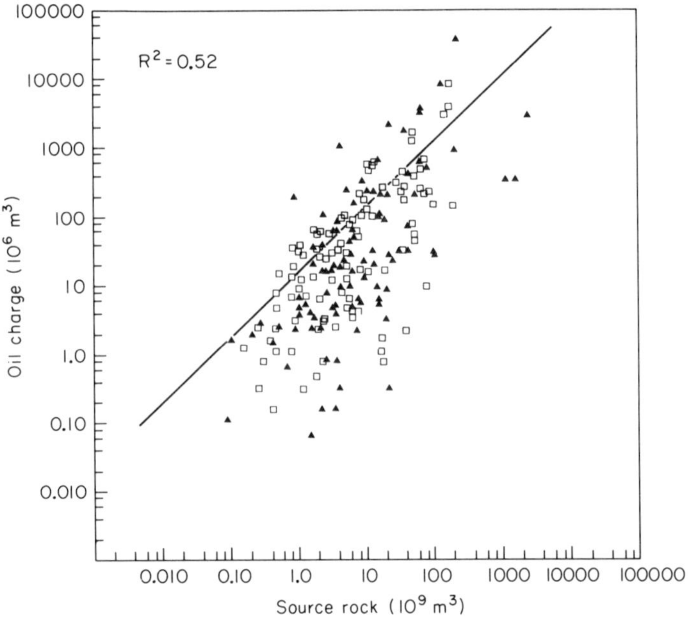

Fig. 3. The same volumes as in Fig. 2 plotted against source rock volume.

and not to the fact that the rich provinces have for some non-technical reason been more densely drilled than the others.

This point is illustrated by Figs 2 and 3. Figure 2 is a chart showing the oil volumes in a number of basins plotted against their sedimentary volumes and it is clear that there is no strong statistical correlation between the two quantities. Figure 3 shows the same oil volumes plotted against one simple geological parameter — source rock volume — which comes much closer to providing a statistical 'explanation' of the basins' richness in oil. This type of observation shows up the weakness of attempts to predict petroleum prospects simply from the fact that few wells have been drilled in an area, or from the existence of basins containing sediments which may or may not be favourable for petroleum. Life is not so simple as that and it is safe to say that most basins which have not yet yielded petroleum after more than a century's existence of the international oil industry are not going to do so now without the application of a carefully thought out and expensive scientific exploration programme.

EXPLORATION MANAGEMENT

Exploration methods and techniques are terms covering an extremely wide and varied range of technical and scientific activities used to discover oil and gas fields. Any good library on the subject would contain at least tens of thousands of items, if not hundreds of thousands, written in many languages. It is therefore not possible to cover the entire field in one talk, and I do not plan to do so. There is an alternative temptation, which is to pick out a few recent research developments with case histories of instances in which they have worked, and to put these forward as the solution to the world's energy shortages. I do not think that this is appropriate at the present stage of development of the petroleum industry. There may have been times in the past when particular technical developments revolutionized exploration. Convincing cases have been made for the contention that such innovations as the introduction of rotary drilling, the torsion balance gravimeter and the reflection seismograph changed the industry almost overnight. Today, however, exploration has become so complex and technically integrated that it is hard to visualize any 'black boxes' which could overturn the established order so rapidly.

I should like, therefore, to put forward a different view of exploration: as an integrated team activity, bringing together the work of experts and specialists of many different disciplines — geologists, geophysicists, petroleum and reservoir engineers, stratigraphers and geochemists, together with a number of people whose flair is for fitting together the pieces of the technical jigsaw. To complete the metaphor, the resulting picture has to be pleasing and convincing enough to attract some investment cash, or exploration will come to a standstill. To hold this view further implies that in some way the 'whole' of the exploration organization is much more than the sum of the parts. It also implies the existence of an exploration community, thinking and working together, such as may be found in a large exploration organization in an international or national oil company. This concept unfortunately undermines the attractive idea of the talented individual oil-finder who is solely responsible for a major new discovery. The talents are still there and will be required more and more in the future, but the complexity of the technical problems now to be overcome has changed the nature of the task from an entrepreneurial to a mainly managerial one. Even so, one of the main managerial tasks is still the recognition and effective deployment of the scarce individuals with the 'flair' referred to above.

THE EXPLORATION MODEL

I described the exploration management task just now as one of allocating scarce resources in some optimal way to exploration activities. This work can be efficiently done only on an exclusive basis and in most countries this exclusive right is in the control of the government. Even to begin exploration involves a contract negotiation with a government in which cash and work commitments by a company are weighed against benefits to be enjoyed in the event of future success. Even at this early stage, therefore, some judgement is required on the resource

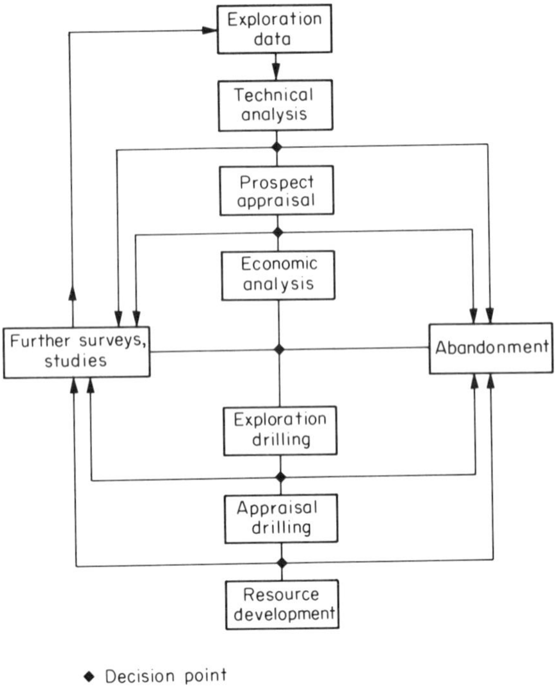

Fig. 4. Simplified model of the exploration management process.

allocation question in terms of balancing real expenditures against hypothetical future rewards.

This process of cost/benefit analysis underlies the very simplified model of the exploration management process shown in Fig. 4. The model is built up in a series of steps. When a new exploration project is under consideration an initial set of exploration data is acquired, either by purchase, from published information or from reconnaissance surveys, and subjected to various forms of technical analysis, the results forming the input of a prospect appraisal. The output of the appraisal consists of a judgement regarding the oil and gas potential of the prospect, either in the form of a qualitative statement or quantified in the form of expectation curves of volumes of recoverable oil and gas. This judgement in turn is entered in an economic analysis in which the estimated exploration costs of an initial evaluation of the area are an important further component. This analysis provides a financial foundation on which to base an investment decision: to commit funds to the project, or to reject the opportunity. In greater detail, the various loops in the model structure illustrate the possibilities of acquiring more data and/or carrying out further technical analysis as further investment decisions are required in the course of the project.

This model structure makes a distinction between the acquisition of exploration data which are of a 'primary' nature (e.g. seismic and well log data) and the subsequent technical analysis, leading to prospect appraisal. The primary data are obtained in the field using capital-intensive equipment, and their acquisition on average accounts for about 80% of total exploration costs. The technical analysis is carried out in the office or laboratory, is knowledge-intensive rather than capital intensive, and much cheaper but certainly no less indispensible.

PRIMARY EXPLORATION DATA

In many ways the scientific and technical effort in exploration over the past few years has concentrated less on data acquisition and more on their treatment and analysis. Table 1 gives a sample list of types of primary exploration data. It is significant that a high proportion of these are normally acquired in contractor-operated activities (e.g. virtually all geophysical surveys, petrophysical well logs) and that such

Table 1
Primary exploration data

Aerial photographs
Airborne magnetometer data
Field geological observations:
hydrocarbon shows
macrofossils
microfossils
rock descriptions
rock sequences
structural information
unconformities
Gravity data
Reflection seismic data
Satellite images
Well information:
hydrocarbon shows
palaeontology
petrophysical logs
sample and core lithology description
structural information
test results
velocity surveys
etc.

items as surface geological surveys are not generally considered to be in the forefront of technology.

Seismic data acquisition is one very important technique in which a large quality gap exists between the average survey carried out by a run-of-the-mill contractor for a moderately aware client, and the best available technology, carefully applied in close co-operation between the contractor and a client fully aware of surface conditions, subsurface lithological layering, structural style and the nature and depth of the prospective targets. This, therefore, is one major area for potential improvements.

Another such area is the observation and description of hydrocarbon shows on well samples. This is an activity of enormous importance for the selection of test intervals and for subsequent prospect analysis. This work is often poorly executed on the rig-site at dead of night, and here again there is considerable room for improvement.

If one looks for new developments in the primary data field the most obvious example is in the growing use of satellite images for primary basin analysis in relatively unmapped or remote onshore areas. Even this, however, has not lived up to its initial promise as a resource-finding tool, and in many instances it is inferior to conventional aerial photographs for petroleum structural mapping.

TECHNICAL ANALYSIS

Table 2 lists some 40 types of analysis or study currently being applied to exploration data, and it is clear that this is a rapidly evolving area of activities since at least about half of these techniques would not have been in frequent use, say, 20 years ago. Table 2 is far from exhaustive and probably understates the variety of disciplines and types of investigation available to a highly qualified and technically aware petroleum exploration organization having its own research facilities. These studies are typically carried out by 'explorers' within the oil company, although a number of them may be available from contractors on a fee basis. As I suggested previously, however, none of these methods is an infallible winner in itself as a means of predicting commercial petroleum deposits. The strength of a successful exploration organization therefore resides in the ability of its staff to combine and integrate many such studies, selecting the relevant and significant results (for selection of a drilling location, for example) and rejecting the remainder.

As a result of this integrated approach there is little doubt that the degree of uncertainty attached to the drilling of certain types of prospect has diminished significantly (in cases where the best exploration technology has been competently applied). Out of the long list of items in Table 2, two groups of techniques have made a particular contribution:

Table 2
Technical analysis

Acoustic impedance studies
Aeromagnetic interpretation
Basin classification
Carbon isotope analysis
Clay mineralogy
Core analysis
Data reformatting/display
Diagenetic study
Direct hydrocarbon indication analysis
DST/FIT analysis
Gravity interpretation
Groundwater geochemistry
Hydrodynamics
Oil/gas analysis
Oil/gas expulsion studies
Oil/gas migration path analysis
Organic metamorphism determination
Palaeoclimatology
Palaeogeography
Palaeogeomorphology
Palaeo-sealevel studies
Paleo-ecology
Petrophysical log processing/analysis
Photogeologic interpretation
Plate tectonic analysis
Porosity analysis/prediction
Radiometric age determination
Regional geology
Satellite image processing and analysis
Seal capacity analysis
Sedimentary environment modelling
Seismic amplitude analysis
Seismic data processing
Seismic 3D modelling
Seismic facies calibration
Seismic/surface geological correlation
Seismic velocity analysis
Source rock/hydrocarbon correlation
Source rock type/richness evaluation
Stratigraphic correlation
Stress field analysis
Structural mapping/cross section construction
Structural style analysis
Subsurface pressure analysis
Subsurface temperature analysis geothermics
Synthetic seismogram construction/analysis
vulcanology

(1) Better definition of structural trap geometry and the configurations of subsurface rock strata, especially in complex areas, as a result of better seismic quality, mainly due to specialized data processing, including migration techniques and model studies.
(2) Greatly improved understanding of the relationships between source rock type and maturity and petroleum occurrence, following the explosive growth of geochemical studies on both the micro and the basin scales.

In addition assistance has been provided in many cases by the increasingly widespread recognition of direct indications of hydrocarbons on seismic sections ('bright spots' etc.), although these are often of value rather as general indicators of the existence of scattered hydrocarbons in the subsurface than as direct indications of commercial oil or gas fields.

These advances are all, however, applicable only to areas for which a certain minimum amount of exploration information is available. This is in contrast to the common situation where all that is available about a particular area is some very sketchy seismic information and the results of one or two old dry holes without modern logs. In these circumstances even the best-equipped exploration organization can find it difficult to make a valid judgement on the prospectivity of the area.

PROSPECT APPRAISAL

In a modern exploration venture the results of detailed analyses and studies can be used to develop quantitative estimates of the amounts of oil and/or gas which may be present in an undrilled prospect, trap, acreage block or basin. From an exploration management viewpoint, the function of such estimates is to improve the quality of investment decision-making by giving a better idea of the chance of success and the range of potential rewards.

The topic of quantitative prospect appraisal is one which has aroused the interest of geologists for many years because of the very large sums of money which are involved in exploration investment decisions. In these circumstances any technique which reduces the investor's risk exposure, even by a small amount, is enormously valuable, and a wide variety of different methods have been developed, ranging from simple subjective estimating procedures to highly sophisticated computer models. However, the best choice of estimating method depends on the amount and quality of exploration data available. In areas for which basic geological information is available prospect appraisal may be based on simple geological models incorporating sizes of traps and extent of potential generating areas. In experienced hands such methods can give very reliable results.

For geological situations in which a greater degree of information is available, Shell geologists have developed an appraisal method based on a computer model simulating the geological subsurface processes of generation, migration, accumulation and retention of oil and gas. Figure 5 summarizes the main steps in the model, a detailed account of which was presented by Dr M.H. Nederlof at the 1980 World Petroleum Congress.

The model is based on a very large data bank consisting of some 500 well-documented oil and gas fields and conclusively-tested dry traps located all over the world, in exploration ventures operated or studied by Shell geologists. The model uses this data bank as a set of world-wide background assumptions which are 'updated' with the specific information on the project under investigation. The project geologists and geophysicists are required to present the results of their technical analysis in the form of precise judgements relating to the generation of petroleum (yes/no?, and how much?) and to its migration and retention in the trap. The simulation of these processes in the model, shown schematically in Fig. 5, is based on Monte Carlo statistical techniques. The result of the simulation is a set of expectation curves of recoverable oil and gas showing minimum recoverable volumes at various levels of probability.

The computer simulation described above is a complex process, but

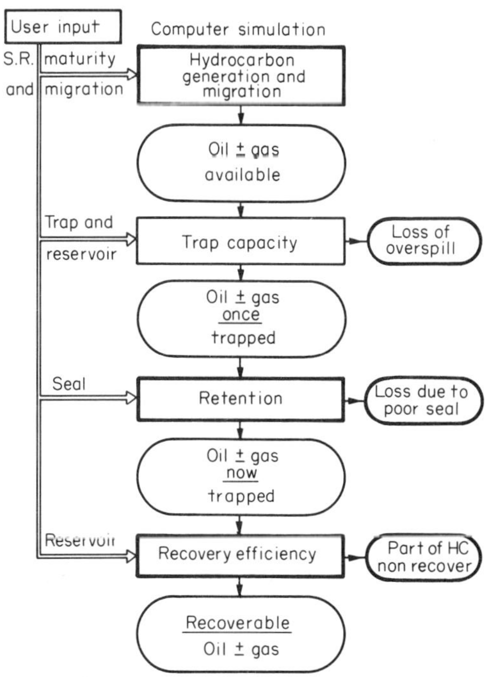

Fig. 5. Simplified flow chart of the simulation procedure.

much less so than the natural geological processes which are being simulated, and which are not yet perfectly understood by scientists. This type of modelling can therefore not be more than an approximation to reality, and its validity depends entirely on the correctness of the geological theories on which it is based. However, from a practical viewpoint the model has been extensively tested and the predicted results are closer to actual drilling results than those obtained by any other method known to us. The technique is widely used in Shell companies as an aid to exploration investment decisions.

EXPLORATION COSTS

Exploration ventures present particular problems of cost management since the costs themselves are very high and unpredictable, the outcome is totally uncertain, and the benefit to the company from spending the marginal dollar is difficult to determine, or non-existent. Governments, on the other hand, often have very rigid views about the size and cost of exploration work programmes they wish to see carried out (at the company's expense!) in geologically highly risky areas. This insistence may deter many companies from entering the more risky exploration projects, and in such cases governments could actually accelerate the exploration of these speculative areas by moderating their terms (e.g. by providing for seismic option contracts without drilling commitments).

In order to estimate the costs likely to be incurred in future exploration campaigns, the best guide is probably the actual experience of ventures now in progress, or recently concluded. The following are some unit cost figures for various exploration projects which have been extracted from the financial returns of Shell exploration operating companies. The unit costs are broken down under the headings of Data Acquisition Costs and Other Costs. In addition to the unit costs a brief review is given of total Venture Costs of some typical unsuccessful exploration ventures.

All the following cost figures are in 1980 dollars.

Data acquisition costs

Actual costs for exploration data acquisition tend to cover very wide ranges depending on such things as the geographical nature of the terrain (offshore or onshore, low-lying or mountainous, swamp, jungle, desert or cultivated, etc.), on the available infrastructure, and on the complexity of the geology being surveyed or drilled. Variability is generally greater onshore than offshore.

Offshore seismic costs currently fall in a range of \$600 000 to \$1 million per crew-month or \$500 to \$1000 per km surveyed, averaging about \$700.

For *onshore seismic* data the following are some typical costs:

	\$'000 per crew-month	\$'000 per km
Desert areas	400	2.5
Western Europe, populated	350	3.5
Tropical jungle	1000	up to 25

Offshore drilling costs are currently estimated in a range of \$3 to \$5 million per month including ancillary services but excluding overheads. These costs may effectively rise much higher in cases involving mobilization of a rig to drill a small programme of, say, one or two wells in a remote area without infrastructure. In the current tight state of the rig market traditional cost differences between jack-up rigs, semi-subs and drillships have disappeared and at present costs depend more on the specific contract conditions negotiated.

For *onshore drilling* the corresponding figures for a heavy land rig such as might be used for a deep exploration campaign in a new area are in the range of \$1 to \$2 million per string month.

When planning future exploration campaigns thought must be given to the future levels of inflation affecting these unit costs. Over the past two years the cost increases affecting drilling and seismic equipment have considerably exceeded general inflation rates. This is because of the high level of worldwide exploration activity following political developments in Iran and elsewhere. Historically these developments have always been followed within a few years by periods of rapid construction and rig and crew surplus, which have tended to bring contract prices back to more normal levels. On this occasion it seems quite possible that such further developments as petroleum price deregulation in the United States may again increase demand for rigs and seismic crews and keep the market very tight for some years.

Other costs

The costs of all the special studies and technical analyses mentioned in this paper, as well as the share of back-up research necessary to maintain them in action, vary greatly according to the particular technical problems involved. On average, however, these costs, together with purely administrative overheads, amount to about 20% of the data acquisition costs in any given venture.

Venture costs

When considering entirely new exploration programmes in areas or countries without ongoing exploration activity, it may be useful to have some typical figures for the costs of an entire venture including data acquisition studies and prospect appraisal and overheads. The following are total costs incurred by Shell companies within the past few years in six such ventures. The number of exploration wells drilled is given in each case as an indication of the dimensions of the venture. I must assume that other companies have had the same experience.

Venture	Type	No. of wells	Cost (millions of 1980 dollars)
A	Offshore	1	21
B	Onshore, desert	2	22
C	Offshore	9	79
D	Onshore, desert	5	93
E	Onshore/offshore Africa	4	32
F	Onshore Latin America	1	7

These very high venture costs are, in effect, the cost of testing the validity of a set of geological concepts. In all the cases cited above, the testing proved to be destructive since the ventures were unsuccessful and had to be abandoned, and all the money invested in them was therefore lost.

CONCLUSIONS

1. The theme of this meeting is petroleum exploration strategies in the developing countries. The successful strategy is one which discovers new petroleum resources.

2. This involves selection and application of the most suitable scientific and technical exploration methods for the area being investigated. The activity of exploration management involves bringing together experts and oil-finders belonging to many different technical disciplines in order to plan and execute the most appropriate exploration programmes.

3. Countries currently importing petroleum are largely those in which exploration has so far been unsuccessful because of unfavourable geology or difficult technical conditions. Successful exploration in these countries therefore demands the very best technology, the most experienced people and the most careful exploration management.

4. Such exploration is extremely costly in terms both of cash and of the skilled exploration manpower required.

5. It is also extremely risky, since the unsuccessful explorer stands to lose all his investment.

6. Those companies having the technical and financial resources to carry out high-risk exploration have to choose their investments based on careful weighing of the balance between risk exposure and the prospective benefits in the event of success. Government strategies can encourage companies to invest in high-risk areas in two main ways:

(a) Recognizing the existence of geological risk by negotiating flexible working arrangements allowing companies to limit their commitments if results are unfavourable.

(b) Providing fiscal and other arrangements allowing the risk-takers an adequate reward for the successful venture.

I would conclude by expressing the view that this is the point where the interests of governments wishing to develop their petroleum resources and those of companies as willing risk-takers have a great deal in common. Petroleum exploration strategies should take account of this commonality of interest.

6

AVAILABILITY OF PETROLEUM EXPLORATION TECHNOLOGIES: A DEVELOPED COUNTRY VIEW

Jan-Olaf Willums

Head of Corporate Planning, Volvo Energy Corporation, Oslo, Norway

INTRODUCTION

The traditional outside view of the petroleum industry is that the major oil companies possess and control the full scope of technologies required to explore for, develop and produce oil and gas, that this technology is very sophisticated, and that the selective application of this know-how has impeded the development of hydrocarbon resources in the third world. This is only part of the reality. Petroleum exploration technology is rather a combination of professional knowledge in petroleum-related disciplines, tools and techniques needed to achieve specific ends, and operational know-how on how to use these techniques. In addition, and maybe most important, is the management expertise to co-ordinate and guide the know-how and equipment in a way to produce actual results; that is, finding and producing oil and gas.

The strength and importance of the international oil company lies rather in its ability to provide the last-mentioned service. The complexity of managing all the individual tasks and services, provided by a host of consultants, subcontractors and in-house experts, arranging the financial and legal issues, all within an over-all economic efficiency requirement — all this requires an expertise which these companies have built up over many years of practical experience. It is this expertise that developing countries need access to in order to speed up the exploration efforts in their part of the world.

Let us analyse the technology and expertise needed in exploring for oil and gas and let us see how third world countries can gain access to it. There are four aspects to exploration:

(1) mapping of resource potentials;
(2) exploration drilling;
(3) evaluation and feasibility studies;
(4) organizational policy and training issues.

MAPPING OF RESOURCE POTENTIAL

According to a recent World Bank study over 40 countries in the third world show a promising potential for oil and gas as judged from a geological point of view. But a lucky combination of several factors is needed in order for oil and gas to accumulate: there must have been a source where the hydrocarbon could develop (source rock); the temperature and pressure must have been just right to allow the hydrocarbons to mature; these must have had a possibility to migrate to a spot where they could be trapped by geological formations, and in a reservoir rock that makes it possible for us today to extract them. All this requires that we learn as much as possible about the surface geology and stratigraphy. After more general geological studies of the region to determine whether there is any reason to believe in the possible existence of hydrocarbons, the most important task is to make a geophysical survey, which relies on highly advanced technological equipment. Today such specialized technologies are operated and owned by service companies that work worldwide and are willing to make their expertise available to oil companies and governments alike. For a third world country there are several ways to initiate such efforts and obtain the results from them.

Untied initial work commitments by oil companies

By requiring one or several oil companies to execute a seismic programme as the initial entry commitment into an area, the government will gain access to important geophysical data. This is an easy requirement in an area where the oil companies believe that there is a good chance of finding oil. Recent examples are the geophysical programmes shot in several blocks off the South China coast: here the Chinese Government required any company that wanted to be considered for later exploration licences to participate in geophysical group shoots, and submit its individual evaluation of the data jointly obtained. This requirement was not linked to any guarantee that the company would get a concession. The possibility of getting into the Chinese offshore adventure alone was enough to attract numerous companies to this venture.

Work obligations of foreign companies

The traditional form in which geophysical work is executed is as an initial part of an exploration contract between a government and an oil company. The company will pay for and supervise the programme in those areas of interest to the company. The actual execution will still be contracted out to a geophysical service company. The interpretation will, however, be done by the oil company, which necessarily also retains the ownership of the data, and generally is requested to provide the government with a set of confidential data. This sytem has been traditionally applied in countries where there is little indigenous expertise to manage, control and evaluate a geophysical programme or where funds for such programmes did not exist. However, it leaves the control initially solely to the foreign oil company and prevents operations in areas that for some reason or other are not interesting for the company to enter.

Conditional work programme of foreign companies

A new approach to stimulate interest in areas that would normally be neglected by the major companies has led to the establishment of the International Energy Development Corporation (IEDC). Founded in 1980 by Maurice Strong and managed by experts from both industrialized and third world countries, IEDC aims at initiating geophysical surveys in third world countries at its own risk. Through a professional evaluation and follow-up of such surveys, the group will be able to attract major companies and consortia by offering conditions normally more favourable to developing countries than if they themselves invited such companies. IEDC may be guaranteed a participation in further work or have a guaranteed repayment arrangement with the other companies entering the prospect. In this way IEDC may play an important role as a catalyst without any cost to the government, to create interest in prospective areas in third world countries.

Government sponsorship of geophysical work

In some countries the government will itself take the initiative in contracting with a geophysical service company to carry out a geophysical survey. This arrangement requires that some government agency or a state oil company should have the expertise to evaluate the data collected, as in India or Indonesia, for instance. But this model may also be of interest to countries with only limited professional capacity. Some geophysical companies are already expanding the scope of their expertise to include evaluations of the data they have shot and processed, so that the government can get advice both on the value of the geological prospects and on how to progress further in exploration. If the area surveyed is regarded as promising by the oil industry, the government agencies have in general no problem in selling the data to oil companies, thereby recovering several times the cost of the survey.

International agency sponsorship

The initial geophysical mapping of potential areas is regarded as an important step in getting a country's petroleum activity started. Consequently, both bilateral and multilateral aid programmes may consider supporting such studies if they are carried out by professional companies. Both the World Bank and other United Nations agencies have sponsored such survey programmes. The United Nations Revolving Fund for Natural Resources Exploration also has as one of its objectives to support such data collection projects, both financially and professionally, not only for hydrocarbons but also for geothermal energy. An example of a bilateral project is the assistance of the Norwegian Government to Jamaica, where the Norwegian oil authority assisted in planning and supervising a survey which was executed by the geophysical company of Norway (GECO) and paid for in full by the Norwegian foreign aid programme.

The conclusions arrived at on the basis of the data collected were subsequently presented at international meetings, and a number of oil companies applied for licences in an area that had previously been regarded as not very promising. High quality data reduced the uncertainty in the companies' geological evaluation and indicated a number of interesting possibilities of oil and gas accumulations.

Risk-sharing geophysical ventures

A new approach in geophysical data gathering was the agreement recently signed between Somalia and GECO under which the executing company itself bore the major part of the survey's financial cost. The data were later sold on a commercial basis to interested companies and the profits divided between the service company and the government. Through such an agreement the government can both increase international interest in the country's hydrocarbon potential and make a handsome profit without making any substantial initial investment.

In summary, there are today many different ways to initiate the mapping of a country's hydrocarbon potential. Their choice depends on what competence and financial resources are accessible, and how attractive the area is to oil companies. In all cases the actual technology is obtained from international geophysical service companies on a contract basis. The technology in seismic shooting and data processing is advancing so rapidly that it requires major seismic survey companies to stay at the forefront, driven by a hard and highly professional market to perform better than their competitors — which is to the advantage of the country wishing to use these services.

EXPLORATION DRILLING

In contrast to the relatively modest expenses involved in the mapping stage, the drilling of the structures defined by a geophysical survey requires substantial funds. Drilling programmes may easily run to tens of millions of dollars for offshore operations. The risky nature of the oil and gas business means that it is mainly the large corporations with an ability to spread the risk over several projects which are willing to spend the money for exploration in new areas. So far, neither the World Bank nor the other United Nations agencies have supported drilling efforts. Loans to a government when the likelihood of a positive result is generally far below 10% is regarded as an unacceptable loan burden on these governments.

The technology required during the drilling phase is much more amalgamated than in a geophysical campaign. Besides hiring a drilling contractor, who executes the actual drilling operations, it is necessary to co-ordinate the work of a large number of other subcontractors covering such specialized fields as bottom surveys, well logging, mud-control, supply vessel operations and helicopter services, to name but a few. In each field specialized companies provide these services world-wide. The responsibility for the over-all co-ordination is normally in the hands of an oil company but it is sometimes assigned to an over-all project management consultant.

For a third world country which has completed a successful resource mapping campaign and identified drillable structures, there are fewer approaches to choose from than in the previous stage of exploration.

Operational agreement with an international oil company

The traditional way is to give an oil company consortium a concession or service contract under which the operator is responsible for drilling a number of wells as part of the work programme.

Normally, little technology transfer is involved as part of such an agreement, although some countries insist that contracts include a training clause, under which its nationals will be assigned to work either on the project under negotiation, or on other projects in which the operating company may be engaged.

Technical co-operation agreement

If the country has already built up a national oil company and has some trained professionals, the government may wish to assign the operational responsibility to its own company, provided that an experienced foreign company will act as technical adviser. This professional assistance may then be paid for directly, or be part of the concession agreement under which the foreign company is entitled to a share of the production or of the revenues therefrom. Such technical assistance agreements have been used both in developing countries (India) and in new oil-producing nations in the industrialized world (Norway).

Project management contract

Provided that the country has funds available for its own exploration and has a certain in-house expertise, its national oil company can draw on the services of a project management contractor, who will then take full operational responsibility for co-ordinating the exploration programme. A number of major oil companies are also providing such services on a contract basis to developing countries. Specific technology transfer and training provisions can be included in such contracts.

The choice of a contractual relationship in the drilling phase depends mainly on the financial resources the country is willing to provide and the indigenous expertise the country's national oil company controls.

FEASIBILITY AND FOLLOW-UP STUDIES

Once the drilling programme has resulted in indications or shows of hydrocarbons, evaluating the physical and chemical nature of the find and studying its commercial possibilities is a task for petroleum engineers, field geologists and petroleum economists. If a discovery has been made under a traditional concession agreement, the foreign operator will have the responsibility for such studies. However, the national oil company should be in a position to produce similar studies, or at least evaluate the relevance of the foreign company's studies as seen from the country's own viewpoint.

Both the World Bank and the UNDP system (*inter alia*, through its Revolving Fund) are willing to provide technical and professional assistance in this phase. Other organizations and companies like OILDECO, the Norwegian petroleum industry development company, are set up for the specific purpose of assisting countries at this stage. An example is OILDECO's work in India, where the potential development of a deep water field off the country's east coast was studied on behalf of ONGC.

ORGANIZATIONAL AND POLICY ISSUES AND TRAINING

All of the above phases of exploration for hydrocarbons rely on a certain professional infrastructure in the country's petroleum sector. The efficient organization of a national petroleum company is mandatory if a technological competence is to be built up. Petroleum research and development and training programmes must be established if this expertise is to be developed further. Competence in energy policy matters and legal disciplines is required if a country is to have any hope of establishing an equitable relationship with foreign oil companies.

Over the last few years international organizations such as UNDP and bilateral agencies have seen the overriding importance of these issues if the petroleum activities in third world countries are to be increased in the long run. The UNDP-sponsored project of establishing the Bangladesh Petroleum Institute carried out by OILDECO is one case in point.

The assistance provided by Norway to both Jamaica and China in policy and legal matters may indicate the possibility of a direct relationship between industrial and developing countries in the transfer of this aspect of technology and know-how.

Something which is just as important also is co-operation among developing countries themselves, which has been supported by bodies like the ASEAN Council on Petroleum (ASCOPE), and also among producing countries, of which OAPEC's efforts in support of training in the Arab world are a good example.

CONCLUSION

Exploration technology is not a single type of expertise concentrated in a single business entity — the multinational oil company. It covers many disciplines and comes from many sources. The management and co-ordination of these diverse sources has been the main strength of the oil company. It is this expertise the developing countries need to acquire or borrow from the foreign companies. In addition, the developing countries need to build up a petroleum sector infrastructure that allows them to deal successfully with foreign companies and accumulate the know-how they are exposed to.

Various United Nations organizations, including the World Bank, have the possibility of assisting third world countries in the latter task. The actual access to technology must be provided by the companies actively engaged in the petroleum industry. Some new thoughts and ideas in this respect are emerging. National efforts like OILDECO and international efforts like IEDC are two such new concepts that might stimulate exploration activity. Some rethinking among the oil companies themselves may now be required, and several oil companies are now looking into new ways to structure the relations between national governments and international energy companies. By introducing such new and creative concepts into the petroleum sector, I believe that the developing countries' access to exploration technology and expertise can be improved and the exploration activity in major parts of the third world can be expanded, to the benefit of both the developing countries and the industrialized world.

7

TRANSFER OF PETROLEUM EXPLORATION TECHNOLOGIES: VIEWS OF THE DEVELOPING COUNTRIES

D. H. N. Alleyne
UN Petroleum Adviser to the Government of Kenya

INTRODUCTION

Attention has been drawn from time to time to the disparity between developed and developing countries in the matter of the total and per capita consumption of energy in general and petroleum in particular. In fact, it has come to be recognized that in international comparisons per capita consumption of energy can, within limits, give as good an indication of economic activity and achieved standard of living as can per capita income data. The reservation stems from the fact that often the major users are also the major wasters of energy.

In the field of discovery and production, the early exploration efforts were either in developed countries themselves — the United States, for example — or in what were outposts of the various empires, whether political or economic. Wherever geographically located, exploration, development and production efforts, especially in the early twentieth century, were undertaken by representatives of what are categorized today as developed countries. The principle of exclusivity was often exercised not only in favour of the metropolis *vis à vis* the colonial or dependent periphery but was even invoked for the purpose of keeping within the 'imperial family' benefits, i.e. petroleum, which should not be exploited by certain 'foreigners'. This can be seen in the All British Policy invoked in, say, Trinidad or the Red Line Agreement, applied in respect of operations in the Middle East, the former Ottoman Empire — both effectively for a time excluding the Americans, who eventually, however, came to play a major role in both spheres, i.e. the Middle East and the Caribbean area.

What was being demonstrated during this early period, however, was the application of progressively advancing and improved technology in the field of petroleum by developed countries in order to promote the discovery and production, refining and transport of petroleum by developed countries, for developed countries. The fact of the location was purely coincidental to the whole purpose and objective.

The post-World-War-II era, however, saw a new development — the advancement to full independence and a status of political sovereignty and control over their natural resources of a steady stream of former colonies or dependent countries, some of which housed geographically the very centres of petroleum industries discovered outside of the developed countries. Examples are evident in the Middle East, the Caribbean, Asia and South East Asia.

The Latin Americans had already achieved political, if not in every case full economic independence, in the early nineteenth century owing mainly to the efforts of Simòn Bolivar, efforts which, however, had led Canning in the mid 1820s to remark, with respect to economic exploitation, 'Spanish America is free, and if we do not mismanage one affair sadly, she is British'.

Mexico in 1938 had nationalized its petroleum industry, thereby incurring the wrath of her great and powerful northern neighbour and other developed countries, which took steps to stifle the new nationalized petroleum industry in the markets of the world. Here then was one developing country that took a brave and bold decision and was then forced by international action to abide by that decision and to undertake its own exploration, production, refining, marketing. There is ample written evidence of the trials, of the problems overcome and of the successes achieved by this developing country in the sphere of activity with which this paper is concerned: petroleum exploration technology and the views of the developing countries thereon. It will be necessary, therefore, to return to the Mexican experience in due course.

Meanwhile, however, it may be useful to observe that many a developing country, perhaps a former colony, wishing to obtain basic geological data about itself would do well to explore the archives of its former masters, be they colonial or economic senior partners or relations. In respect of the Middle East, Latin America and the Caribbean, India, Burma, South East Asia, etc., they should investigate the archives of Britain, France, Holland and the United States. But there is another country, never itself an extra-European colonial master, whose role in the finding and development of petroleum is often completely overlooked. In the archives, both public and private, of Switzerland, in Bern, Zurich, Basel, for example, can be found systematically collected and collated geological data on a number of what are now classified as developing countries. This is because Swiss geologists of high repute (and their numbers have perhaps been disproportionately large in relation to the size and population of their country), impelled by a motive and an imperative not yet fully analysed, took their skills and their capacity for concentrated and indefatigable effort into these foreign territories where their own nation had no particular political connection or obligation. They were in a sense the mercenaries of the world of petroleum geology.

The purpose of this brief review is to show at once the problem and the solution facing the developing countries. They have come to political manhood faced with a situation in which the earlier data pertaining to petroleum geology were to be found in the developed countries. Over the years considerable expertise has been developed, also by the nationals of the developed countries, in the field of exploration.

One has therefore to acknowledge the tremendous contribution made by these technologists and their countries, whatever were their motives or their objectives, in the development of the present world petroleum industry. The question, however, is: where does one go from that point? It was long argued, quite persuasively, by proponents of the developed countries and some institutions largely supported by them that exploration for petroleum is such a risky business that the developing countries should not put their scarce financial resources into this area of activity.

Among the arguments against developing countries engaging in such activities are these:

(1) Exploration is a gamble. The chances of finding crude are particularly slim.
(2) Considerable financial outlay is necessary even for a modest exploration programme.
(3) In order to increase the chances or probability of success it is often necessary to expand the area and increase the intensity of the exploration programme.
(4) Any such programmes, except for the really lucky ones, take a long time to achieve any concrete positive results. Therefore, the total capital investment grows year by year and constitutes a great burden for a poor developing country.
(5) Even when success is achieved, it is after a long period of 'waiting', and in any case exploration is a continuing, almost never-ending exercise requiring ever-increasing sums of money to be spent on testing perhaps less rewarding areas.

The merits and demerits of these arguments will not be debated here. Suffice it to say that for the purposes of this paper, if developing countries are not to engage in exploration activity, then the further question of the transfer of technology in the field of exploration does not arise for them.

It is clear from the insistent demands of the developing countries that they have not accepted these arguments in their entirety. It is not, however, strictly correct to assert, as suggested above, that the transfer question does not arise if the developing countries do not participate directly and fully. The transfer question arises, but only in the weaker sense of the expression, since it involves that arrangement or situation where the technology is employed within the physical boundaries of the particular developing country but the questions of who employs the technology and for whose benefit, and the nature of the benefit, produce answers that do not conform with the ambitions of the present generation of new countries.

If transfer of technology merely means that the developing country does no more than open the door so that a foreign company competent in exploration technology may enter for the purpose of carrying out a programme of exploration, determined by itself (i.e. the company), then there the matter ends.

But, however the matter may be viewed by the developed world, the developing countries have set up a loud and insistent demand for the transfer of technology — in this case, in the field of exploration.

One is therefore forced to inquire into the meaning of the expression 'transfer of technology'. If, as in the situation described above, it implies the application in territory or country A by the representatives of country or area B of a particular technological skill and expertise, then the developing countries have in fact enjoyed considerable technological transfer.

They will be able to argue, it is true, that they could not always persuade the developed country or its corporate representatives to apply the technology in those areas, at those times and for the periods that suited the wishes or needs of the developing countries, but that is, perhaps, another question.

The developing countries are aware of these opportunities for having the technology of petroleum exploration applied within their borders. What they want is to be able to participate in the application of that technology.

One is therefore forced to the conclusion that it is necessary to analyse the nature of technology, to see whether it is at all transferable in any meaningful sense, and what are the terms under which such transfers can be made.

What does the 'transfer of technology' mean? To hear the developing countries pleading for or demanding it, and the developed countries or their companies arguing about the difficulty of transferring it, or bargaining for a quid pro quo, it is possible to get the impression that technology is a fixed pool of something which can be packaged and transported from one place to another. But technology is not in any sense a pool or sum of anything finite in quantity or quality. It is being constantly changed and updated. Also, it is not one thing or one kind of thing. It may involve equipment and tools that can be transferred, but it is more than this. It may involve patents and processes. There are internationally recognized procedures for registering ownership of and rights to the benefits of patents in respect to equipment or processes, and they determine the duration of the right. It is possible to buy the right to use a particular thing or process. But technology is not only processes and patents, however important these are in the concept of technology.

Technology is really a combination of factors that includes tools, equipment, processes and patents. But it also involves a certain intangible additional input, i.e. a knowledge of and an ability to manipulate and use those tools, equipment or processes for the attainment of specific ends: the production of certain goods and services. This involves human development, a conditioning of the human mind to comprehend certain techniques and relationships, how certain things 'work', and why or why not. In fact, petroleum exploration technology may be defined as systematic and formulated knowledge about the industrial art in the field of petroleum exploration. This body of knowledge is always changing, being amended, increased and improved by the research, thoughts, and chance encounters of people within the industry or outside it.

The technology can be learned. Learning it can be facilitated by different forms of assistance, teaching, training, and exposure to it. It is doubtful, however, whether it is possible to 'transfer' it.

What is of real significance for the developing world must necessarily be the ability not only to grasp the know-how in a particular operation but to be able intelligently to participate in the 'rearrangement of prior knowledge', and this involves having persons 'skilled in the prior art'.

Grasping this point is fundamental to an understanding of the problem facing the developing countries. It can provide the key to its solution and can indicate one aspect of the role of their state petroleum enterprises in the transfer of technology — the process of technological innovation. This process involves borrowing, adapting, and experimenting with ideas already present in our fund of scientific knowledge.

The course to be followed by the developing countries is, therefore, the creation within each country of an atmosphere conducive to the development of a fund of scientific knowledge and of persons skilled in the prior art, who, through their state petroleum enterprise, are given the opportunity to borrow, adapt and, above all, to experiment with ideas until these become part of the economically usable technology.

Given the nature of technology and the way it is developed, the term 'transfer of technology' is based upon a misconception. It will be more meaningful for the developing countries and their state petroleum enterprises if, instead of asking for the transfer of technology, they adjust their thinking to comprehend the nature of the problem and seek: (1) to acquire the right to use certain aspects of the relevant technology; (2) to develop within their own spheres of operation the atmosphere conducive to a familiarization with the basic scientific and technical ideas and operations relevant to the industry; (3) the training of their own personnel in the prior art and the development of knowledge of the industry.

The last two of the above developments are crucial. They are essential for obtaining the optimum benefit from the right to use aspects of technology, for, while it is possible to acquire both the patent and the services of imported personnel to exploit the patent through various arrangements including joint ventures, production sharing and service contracts, unless local personnel and the right atmosphere are developed, the benefits to the developing country will cease with the term of the contract of acquisition.

Reference has been made to the possible development of an adversary relationship between the economic power of the transnational corporations and the political sovereignty of the host countries. In fact, most developed countries and transnational corporations and the specialist companies are prepared to make the technology available — but at a price. There is the feeling that since developing the technology in many cases involved a long period of time and considerable expenditure, the technology should not be freely available. It should be exchanged for a fee, or be used as a bartering counter for certain rights to natural resources that developing countries own and control but cannot develop without this key, i.e. the technology.

It must be clear to the developing world that, as far as the developed countries are concerned, the transfer of technology is not unidirectional. There is a quid pro quo. In return for the transfer of technology, there must be a fee or an opening of markets for investment possibilities. The governments of developed countries are prepared to help persuade companies to arrange the transfer of technology, but circumstances must be right and will depend on the existence of favourable conditions in the recipient country, such as agreements on the encouragement and reciprocal protection of investment.

One of the obstacles to the transfer or the acquisition of technology is the price demanded by the developed world and its corporations for making the technology available. The insistent request from the developing world for its transfer must be seen as implying that, in some sense, the price is too high.

The non-cash aspect of the quid pro quo, however, raises a more fundamental problem for developing countries. They see the insistence by transnationals and the corporations' home countries on direct participation in the developing economies as an attempt to continue the old domination under a new guise: as an attempt to deprive them *de facto* of a sovereignty recognized *de jure*; as a reversal, in fact, of their efforts to achieve political and economic independence and control over their natural resources.

A number of steps by both developing and developed countries seem therefore to be necessary. Developing countries have to be prepared to pay something for the acquisition of the right to use processes developed by others. More importantly, they must train their people, and acclimatize them to the new intellectual environment. They must produce personnel with training, experience and familiarity with the natural sciences and, in particular, with the petroleum industry. Only in this way can they ensure that the little they concede will sooner or later win them increased self-reliance in technology. Without trained personnel, they will not be able to use the technology effectively. The developed countries must realize that any concessions they make as to levels of fees or the size and nature of the non-cash quid pro quo can be a profitable investment in terms of goodwill and access to a market whose growth they foster.

The steps outlined above could be facilitated by reducing corporate secrecy, and by a conscious effort to lift the veil and the mystique that surrounds the petroleum industry, for it has been correctly remarked that knowledge about oil technology and oil policies is not as esoteric as the corporate stewardship would have the citizenry believe.

Something that must be considered, however, is the time constraints on the acquisition of technology. The developing world is in a hurry. The present energy crisis characterized by ever-increasing petroleum prices and balance-of-payments problems has only intensified the need of the developing countries to acquire exploration technology and ensure its application within their countries. They wish to compress into one decade or less the technological advances and adjustments accomplished by others in half a century or more. With such objectives, it is necessary to commit resources and energies to the means required. Peoples or nations ambitious for advancement, and in a hurry to achieve, cannot demand that such advances and achievements be handed to them on a platter. They must consciously prepare themselves for the objectives they set for themselves.

WHO OWNS WHAT

A first step for the developing country and its state petroleum enterprise is to discover which organizations or companies own or control various aspects of the petroleum technology. These institutions include the major companies, specialist companies, research firms, equipment manufacturers, private consultants, universities, and technical institutes and schools. The United Nations may be able to play a useful role in starting the collection of those data on an objective basis and without favouring any particular institution or country; but initiative must be shown, and it must be taken by the developing countries themselves.

It is evident that the technology does not 'reside' with one company or a group of companies such as the Seven Sisters, the group of transnational petroleum corporations with which the developing countries and their state petroleum enterprises have normally to bargain or negotiate over national control of petroleum resources. In fact, very few processes of interest to a developing country's state petroleum enterprise are owned or controlled exclusively by a major oil company or an independent firm. Such processes, techniques, or patents are usually owned by specialist companies. In most cases the majors themselves do not have an in-house capacity to do these tasks and so contract the services of the specialist firms for those areas of technology. In the field of exploration, for example, the processes and techniques are owned by a few specialist firms that undertake aeromagnetic and seismic surveys, make aerial surveys and map areas, make soundings and interpret data. There are some small specialist firms that excel in interpretation of seismic data. Some majors have developed and own some of the technology in this area, especially in the interpretation of survey data, largely because they wish to keep secret the results of surveys and tests so that they can capitalize on any successes in initial prospecting by taking up acreage adjacent to a successful find.

Basically, however, the problem for a number of developing countries is that they do not possess persons or a large enough number of persons who, on account of their training, would benefit from associating with these specialist firms during a contract period. This is the fundamental problem that must be remedied. Thereafter, no matter who owns the technology, the developing country will be able to acquire it.

COMPARATIVE EXPLORATION ACTIVITY

Reference has already been made to the disproportion in the levels of energy consumption between the developed and the developing countries. The field of petroleum exploration shows an equally distressing picture in terms of the disappointing level of exploration activity in the developing world. This picture was presented at the United Nations Symposium on State Petroleum Enterprises in Developing Countries, held in Vienna on 7 - 16 March 1978. The present accelerated exploration and drilling activity undertaken in the United States in an effort to reduce the country's dependence on imported energy has further aggravated the situation.

That presentation, among others, forced the African delegates to the United Nations Symposium to hold a series of urgent meetings and to come up with their own joint observations and recommendations for co-operation among developing countries in Africa to remedy a problem described in their document in the following terms:

> On the basis of 1975 statistical information concerning world exploration programmes, only 3% of that effort had been put in the 80 non-producing developing countries, most of which are in Africa. Their conclusion is, therefore, not that Africa does not have oil, but that there is too little effort being made to explore for it. Africa could be harbouring as much oil as the big producer today.[1]

Today, in 1981, the list of 80 non-producing countries has been somewhat reduced. Successes have been scored in a few countries, including the Ivory Coast, the Sudan and Cameroon. However, the picture remains gloomy and indicates the continued relevance of the views of the developing world on exploration effort and exploration technology and their response to what has been referred to as the need for the transfer of that technology to the developing countries.

It has already been pointed out that except in the very loosest sense the expression 'transfer of technology' is based on a misconception. The only meaningful sense is that which in its application is least acceptable to the developing countries, i.e. where foreign expertise decides where and when and on what terms it suits the oil companies to apply that technology and financial capacity to the search for hydrocarbons in a given developing country. Unless terms are right and the relative rewards acceptable compared with the returns involved in placing that exploration effort elsewhere, the foreign corporation will not act.

Of course there are other alternative ways of ensuring the application of exploration technology in a particular developing country. The relative merits and demerits of exploration licences, and concessions, joint ventures, production sharing and service contracts are the subject of other presentations and are therefore not discussed here.

All these have been employed by developing countries over the past decades. This, however, has not prevented the continued existence of that stubborn remnant of 80 or so developing countries referred to earlier where only 3% of the world's exploration effort is taking place.

What this seems to suggest is that what developing countries, and especially those in Africa, are seeking when they speak of the transfer of petroleum exploration technology is the acquisition of the technological and financial capacity so that they themselves may undertake the exploration activity where and when it is dictated by their own national interests. It is this aspect of the matter that must now therefore receive attention.

In examining this question of the views of the developing countries on the transfer of petroleum exploration technology it is necessary to ask oneself whether all developing countries are in a single category. If not, what are the categories?

First, what is the purpose of the exploration effort? It is the acquisition of knowledge of a very special kind — knowledge about the nature of the sub-surface geology in a particular area and its potential for containing or yielding hydrocarbons of a quality and recoverable quantity that would at least repay the cost of the initial exploration and subsequent development.

But to be able to acquire this knowledge, i.e. knowledge of facts, it is necessary to employ another type of knowledge — a knowledge of how and of why, i.e. know-how in relation to the techniques, tools, etc. to be applied in the search and assessment of the data derived from the search and knowing why certain things happen or occur and therefore why certain applications of knowledge are necessary and likely to bear fruit — in the form of geological data, i.e. hydrocarbons.

Some developing countries may be interested only in finding hydrocarbons, or rather in ensuring that hydrocarbons are discovered under the sub-surface of their land mass and adjacent marine areas. If that is the limit of their interest, then all they want in terms of the transfer of technology is to ensure that someone, whoever he may be, applies known technology, i.e. the know-how and the tools, physically within their borders and produces hydrocarbons and provides their country with certain fiscal and physical rewards, i.e. revenues and oil and gas supplies.

This category of country may embrace or include most developing countries at an early stage of their political emancipation and of their economic and social development. The general level of education and literacy may still be low — especially in the scientific or technical spheres. There is a relatively small cadre of professional technical personnel of whom it can be said that they are 'trained in the prior art' and therefore possess the basic knowledge, training and intellectual and industrial discipline that will permit them to participate in the technological processes involved in petroleum exploration. If they cannot so participate, they may have to be content with the role of being mere spectators (at least initially) while foreign experts apply the relevant technologies in the search for petroleum in their country.

There is another (a middle) category of developing countries. These countries are characterized by a relatively literate population; institutions of higher education which train personnel to university level in the physical sciences and engineering (civil, mechanical, chemical) and technicians of a relatively high level of competence.

There exists a relatively modern industrial sector which has enabled a small section of the working population to become acquainted with modern industrial practices and ways of thinking and to acquire familiarity with modern machinery.

While yet classified as under-developed or developing, such countries, having a relatively modern industrial sector, would already have some financial strength compared with the poorest developing countries. There may even have been some exploration activity already undertaken by one or other of the multinational corporations without covering the ground thoroughly in terms of actual extent or deeply in terms of the lower geological horizons.

The financial standing could be such as to warrant a wish to participate actively in the search for petroleum without depending entirely on the multinational corporation. This does not imply that the country could see itself as yet in the position of undertaking all aspects of petroleum exploration on its own or through the activities of its state petroleum enterprise (if one has already been established). It could, however, enter effectively into joint ventures, production sharing contracts and eventually even service contracts with major foreign oil companies and/or independents.

Such an association provides the opportunity for finding petroleum — if nature has been kind in this respect. This is the initial essential condition. Once petroleum is discovered in a socio-economic environment such as that to be found in this category of developing country, the door is open for the acquisition — sooner or later — of the technologies that occur in a petroleum economy.

The petroleum industry has the potential for stimulating growth and development in other sectors of the economy. It can become, in fact, a leading sector which, through backward and forward linkages, promotes the development of the rest of the economy. Above all it provides an ideal training ground for agile minds which have already been exposed to the allied sciences. It is through these cadres and their association with the experts and other trained personnel actually engaged in the search for, and the development of petroleum resources that the transfer, or rather the acquisition, of technology will take place.

But a caveat needs to be sounded. The finding of petroleum can ruin the country if it becomes an end in itself having as its concomitant attraction the acquisition and lavish spending of large revenues. It can lead to complacency and desertion of traditional values and areas of economic effort (agriculture, for example), since the country can now afford high levels of imports of new luxury goods. All labour can be regarded as servile and fit only for lesser mortals imported or permitted to immigrate solely to fill these roles. In such circumstances the purchase of the technology, its application within the country and even the association of some nationals with the process of application of the technology will not lead to real acquisition on a lasting basis of the technology itself, i.e. the know-how and essentially the know-why of the industrial processes.

One of the reasons for this would be that, as production expands and wealth is created and spread within the economy, there will develop a need for managers, captains of industry, advisers, consultants, all of whom will be richly rewarded, sometimes for effort and knowledge which will be much less and much less sustained than will meet the rigorous requirements of the petroleum industry. There will therefore be a constant migration, a brain-drain, out of petroleum into these other sectors. Unless the leakage is made up rapidly by new entrants who stay within the petroleum industry long enough to acquire not only the academic training but the experience which alone can convert them into seasoned knowledgeable petroleum professionals and technicians with the necessary know-how, there may even be a reversion to a dependence or semi-dependence on foreign talents while the nationals enjoy the flesh-pots.

There is, finally, that category of developing country which appears best fitted to succeed in acquiring technology in the field of petroleum exploration. In fact in some cases theirs is already a success story, or at least a story of potential success. These include: Algeria, India, Mexico and Venezuela. How Iran and Iraq emerge from the present conflict is yet to be seen. Although there are areas of low literacy, there are also institutions of higher learning that produce professional scientific and technically trained personnel in large numbers. In Mexico, particularly, an advanced industrial sector has emerged which has facilitated the development of an intellectual acceptance of industrial practices.

In fact it is true to say that of all the developing countries, Mexico is the one that has gone furthest towards the real acquisition of petroleum technology in all or nearly all fields, including exploration. The very circumstances under which it nationalized the petroleum industry in 1938 and established PEMEX, its state petroleum enterprise, forced the country to depend upon its own resources, for the company not only to survive in a hostile environment but to succeed in its role as oil company, that is to say, as the leading sector or growth point in the national economy and as a school or centre of learning for nationals, not only for the petroleum industry, but also for other related national industries and sectors.

In this matter of self-reliance Mexico is alone among the developing countries to have undertaken with its own technical resources the entire exploration programme which first redressed the balance in the post-nationalization era when production had slipped considerably.

Subsequent exploration efforts, also entirely planned and executed by its own resources, satisfied a domestic need for petroleum that had grown not only on account of a rapid population increase but also on account of increased industrial and social developments and a government policy of promoting the growth of energy consumption by a deliberate policy of restraining price increases.

Finally, the recognition accorded to Mexico as a new petroleum giant with considerable potential for further growth and a capacity for exports both of oil and of gas has come about entirely through its own mastery of petroleum technology in all its aspects.

It is true that in the process PEMEX in particular and Mexico in general have contracted large foreign debts that at one time threatened to embarrass them. Both foreign banks and other financial institutions faced criticism about the prospects of certain developing countries (e.g. Mexico, Brazil) ever being able to meet and satisfy their debt obligations. The eventual success story has changed all that. The problem of finance remains, however, one of the main obstacles to the effective acquisition and application of petroleum exploration technologies in developing countries.

FINANCIAL ASPECTS

The acquisition of petroleum exploration technology requires a period of active involvement in the application of that technology. The best possible locale for such an exercise is in the developing country itself — even though some preliminary experience may be gained from training exercises in other countries. But the planning and implementation of exploration operations require capital. Earlier mention was made of the fact that only some 3% of the present exploration effort was being undertaken in some 80 non-oil-producing developing countries — mostly in Africa. There is, however, another estimate for all of the developing countries taken together which suggests that in the 12 years up to 1990, about \$30-40 billion, or 30%, of total exploration expenditure will be required for exploration and development activities in the developing countries. When one considers that in most of those countries the average national income per head is in the region of \$300–\$400, one is immediately aware of the size of the problem and the futility of discussing the transfer or acquisition of technology in purely theoretical and technical terms without reference to the practical question of finance.

One of the 'easy' solutions for most such poor countries is to enter into an arrangement whereby some multinational or independent oil company undertakes the exploration. The company for its part will only do so if:

(1) It thinks that there is a prospect of success. Apart from the occasional playing of hunches, a company comes to that conclusion only on the basis of some 'findings', however vague or indirectly connected with the particular area under reference they may be.
(2) It receives a concession or a licence with a concession possibility if the anticipated prospect is justified by the initial effort. Such old-style concessions are, however, no longer in vogue among the large majority of developing countries. A developing country may, however, see some merit in treating such an arrangement as an investment, even a loss leader involving only a relatively small part of its territory, in the hope that should petroleum be struck it can then 'improve' the conditions applying to adjacent free areas.
(3) It enters into some kind of joint venture or production sharing or service contract with the state petroleum enterprise of the country in question. The developing country is sometimes in a state of unpreparedness for this kind of association in that its state petroleum enterprise is either not yet in existence or exists only in embryo. Here again something is required to 'attract' the company.

In some cases the company is prepared to enter into one or other of the above-mentioned arrangements on the understanding that it will meet all of the initial costs, with the developing country joining in the capital contribution only after commercial production has been achieved. In other cases the arrangement provides for the company's recovery of its initial costs out of current production if — and obviously only if — the venture is successful.

Any of these arrangements, if implemented, will provide an opportunity for the developing country to have its own personnel gain some experience and training in acquiring exploration technology. Ideally, however, the state petroleum enterprise should be operational and able to play its part in the exploration, watching the interest of its country and giving its personnel practical experience in exploration and, later, development activities.

It is often at the development stage, when the state petroleum enterprise has to contribute its share of the capital required, that the nature of the burden becomes evident. While such matters therefore fall, strictly speaking, outside the scope of this paper, attention is drawn to this fact, since it is but the very next step in the process and its realities must be anticipated when an exploration venture is contemplated or being undertaken.

However, the situation where all planning, implementation and financing is left entirely to the multinational company is becoming a rarer occurrence. Developing countries and their state petroleum enterprises are no longer content to be mere spectators or silent partners while the national resource is being explored and then exploited. To play that role would in a sense be a betrayal of the effort and the victory involved in the United Nations Declarations on Permanent Sovereignty over Natural Resources. Assertive participation in such ventures or sole undertaking of them, even where (as is the case even with the operation of the multinational corporations) a specialist company is contracted for certain aspects of the operation, requires considerable capital expenditure for land operations and perhaps more for marine operations when the development stage is reached.

SOURCES OF FINANCE

Since we are discussing the case of non-oil-producing, poor countries, the possibility of financing from past or current oil production which would apply in, say, the case of an OPEC country must be ruled out. Even among multinational corporations, the era of self-generated cash for new oil search ventures is fading away.

The developing country must therefore be prepared to provide financing either from its capital budget or from loans, internal or external, or from grants or other forms of assistance from other countries or from multinational organizations.

When the state petroleum enterprise is being established, it is essential that the financial provisions be adequate both in terms of authorized and paid-up capital and in terms of freedom within its operations to generate cash, organize its budget and build up reserves. If it is well financed and structured, the state petroleum enterprise can borrow. If not, it will be accepted as a dependable debtor only with the guarantee of its owner, the state. But the state itself, in the case of many developing countries, may already be reaching, if it has not yet exceeded, the limits of its acceptable borrowing capacity according to the standards laid down by the international banking and other financing agencies.

In fact, this question has already been raised in international banking circles, which point to the growing indebtedness of developing countries; the problem they face in meeting their debt obligations; the fact that, as a group, they need to spend between 20 and 40% of their export earnings to service their external debt, and the further estimate that some 25 to 50% of all new borrowing by developing countries will go towards repaying old debts. The present competitive upward movement of interest rates in the developed countries, instituted to attract capital flows and strengthen or stabilize the exchange rate of their currencies, makes borrowing by developing countries even more expensive and onerous.

There is therefore no simple solution. Some borrowing by these countries and their state petroleum enterprises there will have to be.

One possible area from which assistance may be forthcoming is the special international fund proposed by the World Bank for which an initial capital provision of $25 billion has been suggested, with major contributions from the developed countries and OPEC members. Some countries have already indicated support for the scheme. Others have not yet responded. It is at this time doubtful what the final and formal response of the United States will be in the present context of the new administration (1981), which is committed to reducing Federal Government expenditure and slashing aid programmes especially to those international organizations that are judged to be poor supporters of United States policies. An initial reaction seems to be that the United States, usually a major contributor to such international agencies, would not support or participate in the new fund.

After a period when it refused to provide loans for exploration and a subsequent period of considerable hesitancy, the World Bank will now make loans available to developing countries for the purpose of financing exploration. In fact, as indicated above, it is now taking the initiative in this matter. Its requirements are often, however, quite stringent, including a provision about strict commercial operations. In some countries — Mexico is an example — the national policy includes the charging of special, some may say political, prices for certain commodities. In Kenya, to take another example, which is threatened with deforestation from the heavy domestic use of wood fuel in the rural areas and charcoal in the poor urban areas, there is a reasonable argument, given the levels of incomes, for selling kerosene at relatively lower prices — to save the country from becoming a desert through drastic rainfall decline and the loss of tree cover.

In any event, loans once contracted have to be repaid. The recent appeal to developed country creditor nations to waive repayment of loans by developing countries has not received great support from most of the lenders. In fact one wonders whether such an arrangement may not exacerbate the relationship of dependence which is so strongly resented by developing countries.

It is proposed, therefore, to offer a possible solution or contribution to a solution in the terms set out below.

AN EXPLORATION FUND

Most developing countries, in spite of their poverty, consume petroleum products. Some import the final products while others import the crude oil, which is refined for internal distribution and re-export of certain portions of the barrel. Some again import both crude and products as the domestic requirements, the refinery capability and the crude quality permit. Also, they have all been able, often with great difficulty it is true, to survive prices that have increased from $1.56 a barrel in 1968 to $32.00 in 1981.

The proposal is that each such country should establish a fund which would be earmarked for financing petroleum exploration and development. The proceeds may be used directly for such financing or, more likely, for servicing any loans contracted for the purpose of financing such exploration and later development activities.

The fund itself would be serviced by the proceeds of a levy — a petroleum exploration and development levy — falling on each barrel of crude petroleum or gallon of product imported into the country.

The rate of the levy can be varied from time to time, depending on the assessed requirements of the exploration and other efforts contemplated and on the level of imports. Should petroleum be discovered and produced, the levy could be made to apply to oil and gas produced and utilized. Exploration is a continuing business since new discoveries are constantly needed to ensure that, given the decline curve in the production patterns of the wells, the level of national production and reserves is maintained if not improved.

This fund can be the 'seed-corn' of the national effort aimed at creating a petroleum industry and at participating actively in the technology involved. It is a present investment for future self-sufficiency in energy, a possible relief from foreign exchange burdens. It must not be 'raided' for use in any other project save perhaps, *in extremis*, for a national emergency of the highest order.

The creation of such a fund with the characteristics indicated above will be a signal to those at home and in international circles that the country is serious about its exploration effort and about acquiring the relevant technology. Its very presence will be welcomed by prospective creditors and by any specialist companies whose services will be required initially and at subsequent stages of any exploration programme.

REGIONAL CO-OPERATION

What are the prospects for the creation of funds of this nature in a number of developing countries in a particular region? Can these national funds be consolidated into a regional petroleum exploration and development fund wherein each contributing country has an allocation or minimum quota for domestic exploration, research and development plans related to its input?

The majority of delegates at a recent United Nations Symposium[2] agreed, of their own accord, 'to remind their respective governments to expedite action in setting up a regional organization for mutual co-operation and assistance in oil matters among African countries as was recommended by the first African Conference on Hydrocarbons, re-recommended by the second African Meeting on Energy, and endorsed by the fourth Conference of African Ministers.[3]

It is interesting to note the reasons given for undertaking this important step. These included the following:

> Exploration for development of oil resources is a very costly venture, and success of this needs co-operation among those involved in this industry . . . [There is therefore the need to] share in heavy costs, risks, experiences, information and know-how. Interstate co-operation in the oil industry is a must for success.
>
> Regional mutual assistance organizations in oil matters exist in other parts of the world, such as OAPEC in Arab countries, ARPEL in Latin-American countries, South East Asian countries, and even in the Western countries through their transnationals. . . . In Africa, some state petroleum enterprises are already formed and are functioning, some are formed and struggling to find their way to function, some are in formation, and some are yet to be formed but, nevertheless, are being considered. It is, therefore, necessary to start an interstate body to help these governments and state petroleum enterprises.
>
> The United Nations encourages the establishment of such mutual assistance bodies in economic matters. International aid institutions are favourably disposed to giving aid to such regional bodies that co-operate and help member countries and serve whole regions rather than individual countries.
>
> Oil fields and oil bearing formations transcend international boundaries, so it is economically cheaper to carry out certain oil operations, such as seismic and preliminary exploration surveys, on a regional basis rather than one by one in individual countries in that zone.
>
> There are certain problems in oil exploration, exploitation, transportation, and marketing which are peculiar to the African region, and which need joint and co-operative action by all those member countries taking part in the oil industry in the region.[4]

On the specific question of exploration and finance the views of the delegates were:

> There are many small countries in the region where there are possibilities of the existence of oil — at times even in big commercial quantities — but because of lack of capital and know-how, those countries cannot carry out the preliminary seismic and/or exploratory surveys, the results of which can be used to easily attract loans available from international loan institutions to develop these deposits. A regional co-operative body of the type suggested can pool the individual financial resources of the members to hire or contract companies with the know-how to carry out exploratory or seismic surveys in groups of member countries

> on a zonal basis at very much reduced costs. Such a body can afford to hire or establish firms with modern and sophisticated exploration technology to be used by countries starting in oil industries, or anyone else needing such services in the region.[5]

It is regrettable that so far Africa, a continent that has the greatest need for this type of co-operative effort in the field of petroleum, has not been able to make these ideas and exhortations a reality.[6]

The benefit to be derived from a properly organized and operated fund can be considerable. It is not too much to think that once such a fund is established, it will attract contributions not only from member states but from other nations, and from international bodies of all kinds that show an interest in and are sympathetic to the aspirations of Africa.

In terms of disbursements from the fund, not all countries will have identical and simultaneous expenditure requirements. For example, country A may need only $0.5 million in year one, during which country B needs nothing, country C $1 million, etc.

In this way the region as a whole through its joint efforts can finance or underwrite loans for financing the staggered requirements of its individual members.

A regional fund of this nature, however, requires and must be predicated upon mutual trust and goodwill among potential and actual participants. If good neighbourliness is non-existent or cannot be developed, there is no hope of success. But if the countries of Europe that have fought innumerable serious wars over the past centuries, and two major world wars within living memory, can come together not only in a common market but also in an economic community and a European Parliament — notwithstanding occasional bitter bickering and disagreements about the apportionment of burdens and benefits — what better lesson or example is there for the developing world to encourage it to sink its individual differences in the face of economic adversity and co-operate for the common good?

CONCLUSION

Earlier reference was made to certain countries that had succeeded in acquiring petroleum exploration technology and applying it within their borders. Algeria and India undertake approximately 90–95% of their exploration effort. Venezuela and Mexico have achieved 100% of exploration without outside assistance.

But have these examples of success really resulted in the finally satisfying situation for all concerned? What level of acceptance in international petroleum and general economic circles awaits a developing country and its institutions when they have braved and defied the established patterns and succeeded in this area of industrial/economic activity which was once the preserve of the developed countries? Are there other areas or aspects of a supposed inferiority in the minds of both developed and developing countries that present barriers to full acceptance into the club of technologically competent nations and peoples? In considering these questions one could, with profit, consider the following statement made allegedly by a Mexican official after a recent announcement of rich finds of oil and gas in Mexico: 'We would like the United States to see us as an adult country capable of managing our own affairs'.

These words are an eloquent formulation of the views of the developing countries on the transfer of technology, not only in the field of petroleum exploration but in all other areas of economic activity, and on the normal relationships and intercourse between nations and peoples.

It is only in this way, after the many obstacles referred to above have been surmounted, that the developing countries can succeed, not only in acquiring technology, but in solving a much deeper, more powerful problem. It is a problem that is evident in such expressions as:

> The developing countries see in the attitudes of the developed countries an attempt to restore the old domination in a new guise, an attempt to deprive them *de facto* of a sovereignty acknowledged *de jure* and a reversal of their efforts to achieve political and economic independence and control over their natural resources and their destinies.

> There exists an adversary relationship between the economic power of the international corporations and the political sovereignty of most countries.

> 'The burden of resentment' which President José Lopez Portillo of Mexico mentioned to President Jimmy Carter of the United States in Mexico City in February 1979.

Calls for the transfer of technology are not an end in themselves. They are part of a wider concept, a wider objective. They are part of the New International Economic Order. They are part of a struggle for human dignity, for achievement of acknowledged self-respect.

NOTES AND REFERENCES

1. United Nations Centre for Natural Resources, Energy and Transport, *State Petroleum Enterprises in Developing Countries*. Pergamon Press, New York, 1979, p. 105.
2. United Nations Symposium on State Petroleum Enterprises in Developing Countries, held in Vienna, 7-16 March 1978.
3. Recommendations made by the African Delegates to the above Symposium. See *State Petroleum Enterprises in Developing Countries, op. cit.*, pp. 101-105.

4. *Ibid.*, pp. 101-102.
5. *Ibid.*, p. 102.
6. It should be noted also that the delegates, aware of the technological inadequacies of the countries involved, see a major role for the foreign companies in exploration. The domestic participation must, however, be pursued.

8

CONDITIONS FOR ACCELERATING THE TRANSFER OF OIL EXPLORATION TECHNOLOGY TO DEVELOPING COUNTRIES[1]

Secretariat of the United Nations Conference on
Trade and Development

INTRODUCTION

Developing countries are showing increasing interest in the transfer of technology for oil exploration. This is for a variety of reasons. First and foremost, exploration of domestic oil is becoming more and more attractive, in view of international oil scarcity, in those oil-importing countries which have some promise of finding oil.[2] These countries, however, have no oil exploration experience and hence no domestic technological capacity in this area. Transfer of technology is therefore needed for undertaking the exploration activities.

Secondly, the oil-producing countries in the developing world are finding it necessary to build up their technological capacity in the oil sector in order to control and manage increasingly scarce oil resources. It is also important to recognize that for many of these countries the oil sector, being the only industrial sector, has to function not only as an income earner in the short run but also as a seed-bed for the generation of skills and know-how for long-term industrialization.

Although it is recognized that the foreign oil companies will continue to play an important role in oil exploration activities in developing countries, there is also a growing concern on the part of these countries to undertake or to guide increasingly the work of finding and producing oil on a national basis. It should also be mentioned that in developing countries with only small oil deposit prospects, foreign oil companies may not usually find it economically viable to become engaged in the exploration, so that the only way of exploring for oil in these countries is for the countries themselves to take an active role. An effective transfer of technology assumes a critical role in the light of these considerations.

Given this interest in the transfer of technology, the question of availability of petroleum exploration technologies and terms and conditions of access by the developing countries assumes great importance. In 1978 the UNCTAD secretariat prepared a study[3] in which considerable attention was given to the structure of the international market for oil exploration technology, and the experience of developing countries in acquiring such technology from this market. Following this study, an in-depth study is being carried out for the purpose of examining ways and means of accelerating the transfer of oil exploration technologies to developing countries, with a particular focus on alternative channels for such transfer. The present note is based on the findings of some of these studies.

In the next section, the structure of the international market for oil exploration technologies is briefly described in terms of degree of concentration and obstacles to market entry. In the following section, various types of alternative arrangements for the transfer of oil exploration technology to developing countries are identified, and the more recent experience of some of the developing countries in respect to these different arrangements is discussed. The final section discusses the conditions that are pertinent to accelerating the transfer of oil exploration technologies to developing countries.

INTERNATIONAL MARKET FOR OIL EXPLORATION TECHNOLOGY

Petroleum exploration depends on a complex set of science-based operations such as geological surveys, geophysical and seismic studies, logging and drilling. These operations require know-how, special equipment and highly skilled manpower, which together make up what is referred to here as oil exploration technology.

Transactions for the transfer of individual oil exploration technologies usually take place in the form of provision of technical services and provision of oilfield equipment. However, the boundaries of the oil exploration technology market are not very precise. On the one hand, this market is not independent of the oil resource market. On the other hand, many technological transactions take place as part of wider transactions through which oil exploration activities are organized. However, it is possible to identify the main technology holders and estimate their share in the international market in order to form an idea of the degree of seller concentration.

Whereas the degree of concentration will show one aspect of the market structure, the obstacles faced by potential entrants into the market will give a more complete idea of that structure. These obstacles act as a deterrent for a potential entrant to a particular market, and can best be analysed using the concept of barriers to entry.[4] As the discussion in this section will be concentrated only on the supply side of the technology market, the obstacles to which reference will be made are those generally faced by any firm — from a developed or a developing country — that seeks to enter the market as a technology supplier. The firms which are located in developing countries face additional constraints, which will be discussed later.

Degree of concentration

The market for technical services and equipment for oil exploration is subdivided into specialized markets. The suppliers are oil contractors, manufacturers of equipment and materials, and engineering firms.[5] Oil contractors play an increasing role in the world oil technology market. They carry out very specific tasks relating to geology, geophysics, drilling, logging, reservoir engineering, etc. Oil equipment suppliers manufacture and provide the equipment. Usually they are very specialized and, for a given equipment, a few firms generally share the whole market. The engineering companies are responsible for organizing the building of most of the energy installations, including the provision of the required equipment through affiliates. Their object is to supply complete industrial units, beginning with initial studies and involving supervision of installation, including manpower training.[6]

An estimate of the degree of concentration (as shown by the market shares of companies operating at world level) in the major markets for exploration technologies is provided in Table 1. In the case of technical services, some markets are highly concentrated. In electrical logging, one company, Schlumberger, accounts for 84% of the world market. Halliburton has more than half of the world market for cementing and factoring. With regard to oil equipment, the market for some particular items is highly concentrated. This is the case with drilling appliances and pipes, in which a single firm accounts for about 60% of the world market. A similar situation is found in the case of packers, an item in well equipment.

No estimates are available of the world market for engineering services relating to petroleum exploration. Table 2 lists the ten largest energy engineering companies which operate in the oil sector and the approximate values of their projects in hand. According to the projects they accounted for in 1976, the three principal companies — Kellog, Lummus and Stone & Webster — had 55% of the leaders' market.

In addition to the companies listed in Tables 1 and 2, small companies and the transnational oil companies also supply technical services for oil exploration and production. In some local markets, many small-scale firms and consultants are engaged in technical services. In the case of drilling, for example, besides large international contractors like Forex (an affiliate of Schlumberger) and Parker, a multitude of small United States drilling companies, many of them having only one rig, are active in the domestic market.

The oil companies control the know-how for the operations (geophysics, drilling, production methods from fields, etc.) but often do not hold all the relevant technology.[7] They have therefore to play the role

Table 1
Selected major oil contractors and oil equipment suppliers: sales and market shares, 1976 (in millions of dollars and percentages)

Company	Equipment						Technical services					Total sales of companies[a] ($ millions)
	Drilling equipment			Well equipment			Logging		Production-related services			
	Bit	Drill pipes	Join-tures	Other drilling appli-ances	Packers	Security equip-ment	Open hole	Cased hole	Cement-ing	Acidifi-cation	Factor-ing	
Baker International	12	6	31		56	15						553
Dresser Industries	17				12	31	12	16				2232
Halliburton					24	18	2	6	58	47	53	4866
Hughes Tools	38	3	58						10	6	4	383
Schlumberger				10			84	44	13	18	18	1815
Smith International	25	58		60								308
Others	8	33	11	30	8	36	2	34	19	29	25	
Total (industry)	100	100	100	100	100	100	100	100	100	100	100	10152

Sources: *Bulletin de l'industrie pétrolière* (Paris), 27 December 1977; Crédit industriel et commercial (CIC), 'Note d'information économique et financière' (Paris), No. 8, November 1977.

a Figures refer to the total sales of the companies, including sales not accounted for in the market share columns. All the companies except Schlumberger, which is Franco-American, are based in the United States of America.

Table 2
Major energy engineering companies, 1976

Company	Country	Projects in hand, 1976[a] $ million	Projects in hand, 1976[a] Percentage	Sector[b]
Kellog	United States	4225	21.9	O G N Ch C
Lummus	United States	4120	21.4	O G N Ch C
Stone & Webster	United States	2240	11.6	O G N Ch C
Lurgi	Germany, Fed. Rep.	1919	10.0	O G N Ch C
Fluor	United States	1572	8.2	O G Ch
Foster Wheeler	United States	1470	7.6	O G N Ch C
Creusot Loire	France	1068	5.6	O G N C
Badger	United States	1012	5.3	O G Ch
Snam Progetti	Italy	832	4.3	O G N Ch C
Technip	France	795	4.1	O G
All above-mentioned companies		19253	100.0	

Sources: *Chemical Age* (London), June 1974, July 1975, June 1976, July 1977; *L'industrie du pétrole dans le monde gaz-chimie* (Paris), No. 489, 1977, p. 59.

a It should be noted that, respecting the wishes of their clients, many engineering firms are reluctant to quote the value of projects in hand. Their relative positions cannot therefore be reliably established. Nevertheless, almost everyone would agree on the 'top four'.

b O: oil; G: gas; N: nuclear; Ch: chemistry; C: coal.

of 'co-ordinators' of operations and intermediaries between the service companies and the oil-producing countries.

Regarding the degree of concentration in the technical service and oil equipment markets, it may be concluded that dominant positions have been established in certain particular markets, for example, reservoir engineering, electric logging and computer software-using services. In other technical services and oil equipment markets, suppliers are more fragmented.

Obstacles to entry into markets

A review of the obstacles to entry into the markets for oilfield technical services will give an idea of the factors favouring existing firms in this business. Regarding the absolute cost advantages of established firms, the following factors may be mentioned: (a) although proprietary technology is important only in data processing for software and electrical logging, secret equipment design and complex know-how are significant factors in technologies like reservoir engineering, computerized interpretation for drilling control and mud logging; (b) material and equipment are generally available from many competing sources; however, some cost advantages accrue to service firms belonging to groups that supply their own equipment; (c) availability of highly qualified personnel is a great advantage for established firms; (d) managerial capacity is a very important factor, which favours existing suppliers.

Product differentiation as such is not important in this specialized technology. However, the reputation of international suppliers and their ability to provide equipment and services adapted to all kinds of conditions are important factors favouring existing firms.

Economies of scale are a fundamental factor in the international market. This is particularly relevant in respect of the accumulated information and know-how stemming from years of experience and operation in different geographical areas and under different conditions. In addition, by operating on an international scale, a company is able to allocate substantial funds to research and development.

The obstacles to mastering operating technologies are generally minor, except for a few technologies (logging; reservoir engineering; rigs, logging and completion in offshore drilling). The barriers to entry are higher for construction of the specialized equipment needed for oil exploration and production. This is the case with equipment for offshore drilling and for geophysical and seismic work.

Access to the technology for exploration is difficult because of the charactertistics of the economics of exploration and the role played by the transnational oil companies. Oil exploration is a risky activity, since it is not known in advance whether oil will be found, and in what quantity. The random element is very strong in the discovery of a new field. However, it is much smaller once the field is discovered and the development stage begins.[8]

The distribution of the cost of onshore exploration may be roughly broken down as follows: reconnaissance and geological work: 10–20%; geophysical work (mainly seismic): 15–30%; drilling: 50–75%. Offshore exploration is more expensive than onshore; it involves between $20 million and $50 million for an exploration campaign extending from three to ten years. Geological and geophysical work combined usually accounts for 20–30% of offshore exploration costs.[9]

Transnational oil companies have three main advantages over newcomers in oil exploration. First, their functional and geographical diversification enables them to dilute the risk element, which is a major characteristic of oil exploration. Second, they possess data banks which facilitate the interpretation of information obtained in the process of exploration. Third, they are very well acquainted with the oil technology business and have organizational, technical and financial capabilities that enable them to fulfil the role of co-ordinators. In this way, although transnational oil companies are not the major technology owners in exploration, they play a predominant role in this market.

These factors should be taken into account in any analysis of the structure of the world market for oil exploration technologies. The relatively low degree of concentration and low barriers to entry into the many sub-markets into which oil exploration and production technologies are divided open new perspectives for oil producing developing countries in this area. However, in the exploration stage, the capacity to take risks, the importance of the data banks in facilitating the interpretation of the results and the managerial capacity to co-ordinate these expensive operations in an efficient way are factors that should also be taken into account and that, in the meantime, favour the established transnational oil companies.

DIFFERENT ARRANGEMENTS FOR THE TRANSFER OF OIL EXPLORATION TECHNOLOGY: EXPERIENCE OF DEVELOPING COUNTRIES

Broadly speaking, the transfer of oil exploration technology to developing countries has taken place through two different types of arrangements. One type consists of those arrangements which have been negotiated as part of broader oil exploration agreements with international oil companies. Although the predominant concern of the developing countries in negotiating oil exploration contracts has so far been limited to matters such as distribution of exploration risks, division of profits and control over production and physical oil, they seem to be paying increasing attention to the terms relating to technology, particularly in recent years.[10] It may therefore be useful to investigate how technology is transferred through exploration contracts.

The other type of arrangement relates to agreements concluded by developing countries with different technology suppliers for the specific purpose of acquiring the technology or technological capacity for oil exploration. They include commercial agreements for the transfer of specific know-how or the right to use patents, technical co-operation agreements of intergovernmental nature, establishment of subsidiaries of transnational joint ventures between these and domestic firms for the supply of technical services, etc. Whereas the technology transfers concluded as part of exploration contracts are usually specific to the projects covered under the contracts, these technology transfer arrangements have broad application independent of any particular projects and are complementary to technology transferred through exploration contracts.

In the following pages, a brief review is made of the experience of some developing countries in respect of both these types of arrangements to indicate the extent to which they have served in the strengthening of technological capacity for oil exploration in developing countries.

Technology components in oil exploration contracts

Risk contracts

Oil exploration contracts are described by a wide variety of names.

A great majority of the contracts so far concluded by developing countries are the so-called 'risk' contracts under which the oil company puts up all the money for exploration and development and takes all the risk.[11] The company also organizes the exploration work and plays the role of co-ordinator of technical services required for the operation.

In order to obtain some idea of the extent of technology transfer through these risk contracts, a survey has been made of the summaries of over 600 such contracts signed during the 1970s to extract their technology components.[12] The summaries derived in large part from those compiled by Barrows, a New York-based consultant firm specializing in reporting on contracts and legislation in the oil industry. Using these summaries as a data base has certain limitations, including the following: (a) since Barrows' summaries focus more on financial terms than on less readily quantifiable dimensions such as technology transfer, they may not be taken as conclusive evidence on technology transfer terms;[13] (b) because of the time-lag in reporting by Barrows, the data base does not adequately reflect the situation in the more recent period; (c) the data base includes not only developing countries but also some developed countries. Nevertheless, the summaries may provide broad indications on the subject at hand, and it is only in this limited sense that the results, presented below, are to be interpreted.

The survey indicated that of the 600 contract summaries examined, 88 (or roughly 15%) included some specific mention of the foreign partner's obligation concerning technology transfer. There was evidence of increased attention being given to technology transfer terms in exploration contracts in the second half of the 1970s (23%) compared to the first half (10%).

The technology transfer terms found could be classified into the following five types: (a) training requirements; (b) provision of technical requirements; (c) provision of scholarships; (d) employment stipulations; (e) other requirements relating to technology.

'Training requirements' were the most common technology transfer component in the risk contracts. For the 1976–1979 period, for example, over one-third of all the technology transfer terms included in the contracts related to these requirements. It is also important to note that there was a qualitative change in the specification of these requirements. In the early 1970s, the terms related broadly to general requirements to train nationals; in the second half of the decade, however, they came to include specific types of training to be imparted by the oil companies as well as specific requirements for the replacement of expatriates.

'Provision of technical requirements' (i.e. a term which obliges the company to provide hard- and software, personnel and services) was mentioned in some 17 contracts signed in 1974 or later. Only two out of these 17 contracts explicitly indicated that the technology so provided would be absorbed and indigenized. In the absence of such specification, this term may mean that the company is to do the operational work as is expected under most risk contracts without reference to absorption of the technology involved.

'Employment stipulations' concerning the replacement of expatriates by nationals were mentioned in 22 contracts, all signed in 1974 or later. However, these terms were usually general invocations to use indigenous personnel 'where available' or 'if qualified' in the judgement of the companies. There was also no apparent trend for these terms to gain in specificity or to become more obligatory.

'Provision of scholarships' was included in only nine of the contract summaries examined. Except for some which indicated the amount of money to be spent, and required the scholarship programme to double or triple in size after the company had made a commercial discovery, they appeared to receive little attention. There was no discernible trend for 'other technology-related terms'.

It may be concluded from the above that, on the whole, risk contracts served as a relatively poor vehicle for technology transfer during the 1970s. This conclusion will be further strengthened if one takes into consideration the likely possibility that some of the contract provisions may remain unimplemented. It should be stressed, on the other hand, that there was an apparent trend for the technology transfer terms to gain in relative importance over time, and to contain quite specific references. This is significant because it coincided with the general trend for the governments of developing countries to secure a greater share of profits, to exert more control over the pace of production and to retain larger proportions of physical oil.

Some developing countries have also demonstrated that a long-term plan for transferring the technological capacity in the oil sector can be woven into risk contracts through a deliberate government policy.[14] Thus, terms may be negotiated to maximize openings for the training of nationals and for staffing key positions with graduates of such training programmes. Joint management arrangements with the company may be made to oversee exploration and development work through which domestic capacity to control and manage the oil sector projects may be gradually developed. The wide variety in the technology components of the risk contract itself may be a sign that much improvement can be made to increase their usefulness as arrangements for technology transfer.

Service contracts

Service contracts for oil exploration are used by countries which carry out their own exploration, taking the risk alone, but which need to contract the international companies or independent suppliers for provision of technical services. The principal characteristic of these contracts is that the country concerned retains all of the profits, controls the disposition of oil, and also retains the option of controlling the production operation itself.

The main technological implications of service contracts are as follows:

(a) Access to technology is not complicated by the financial arrangement for the exploration (and it is important to remember that, as mentioned in the preceding section, the market for oil exploration technology is not necessarily a monopolistic one and that the suppliers are often numerous and diverse, especially for onshore or shallow water exploration).

(b) There is scope for providing technology from the domestic sources if they are capable of providing it, thereby gaining opportunities for learning, although a 'price' may have to be paid in terms of performance, at least in the short run.

(c) Negotiation and execution of a service contract itself requires a considerable amount of managerial and organizational capability as well as of knowledge concerning the technology involved and its market.

However, the signing of service contracts with the technology suppliers does not necessarily mean transfer and absorption of oil exploration technology by the country concerned. In fact, in many oil-producing developing countries only the 'leasing' of technology takes place for a vast array of oil-related services, including not only exploration and production but also refining of crude, and marketing arrangements. In such cases, not only are opportunities for technological learning foregone but often very high prices are paid for the services.[15]

Some developing countries, however, have attempted to combine service contracts with technological development effort in the oil sector. A good example of this is found in the experience of India in offshore exploration. Given the lack of experience in this advanced technology and the need for rapid exploitation of the resources, its Oil and Natural Gas Commission (ONGC) reached an agreement with foreign contractors to subcontract work to a local manufacturer for the provision of the equipment, it being understood that the foreign firm should be fully responsible for the quality and delivery schedule of the entire system. At the same time, ONGC was to develop in-house capability in the matter and to retain an external consultant to check the work done by itself and by the Indian firm.[16] By combining these efforts with foreign technical co-operation through public and semi-public channels and, most important of all, indigenous research and development effort, India could build up a solid domestic capability in the offshore technology.[17] In this way the 'do-it-yourself' approach to oil exploration implicit in service contracts may be extended to encompass national technological capabilities.

Technology transfer agreements

In discussing the second type of technology transfer arrangements — that is, special arrangements made for the transfer of technology — it is useful to distinguish between two categories. One is the establishment of subsidiaries by the oil companies and transnational service companies for the supply of technical services, and the other the conclusion of technical collaboration agreements by the national oil company with the foreign technology suppliers.

Subsidiaries of oil companies and transnational service companies such as those listed in Table 1 are actively engaged in the provision of technical services in most of the oil-exporting developing countries. They receive the technology from their parent companies, import the capital equipment needed to carry out exploration, and train some indigenous personnel. They play a predominant role in many individual oil service sub-markets even in countries like Venezuela where some indigenous technological capacity has been built.[18] Their activities, in effect, represented the realization of the technology components in the exploration contracts discussed above; and how much they actually contributed to the strengthening of the technological capacity in the exploration field in a particular country where they operated depended primarily on the extent to which the relevant technical skills were imparted to the local personnel through training.

In those developing countries where a certain amount of technological experience has been accumulated in the oil sector, technical collaboration agreements with foreign technology suppliers have been used for either technical assistance or licensing of specific patents or know-how. In some cases, they represented technological agreements signed with the former operating companies following nationalization in order to ensure technological continuity; this was the case in Iraq and Venezuela. In others, as in Mexico and India, a domestic technological capacity in exploration was gradually developed utilizing these technology transfer agreements in combination with other arrangements.

Joint ventures with the technical service suppliers were also established in a number of developing countries, particularly in those areas where the gap between the domestic and the foreign technology was considerable, and where the shortage of domestic manpower resources was acute. Such arrangements represented a formula combining the two above-mentioned approaches. The most notable example of this has been provided by Algeria, where SONATRACH, an independent state corporation, entered into several joint ventures with specialized service companies to supply such services as drilling, geological surveys, well-logging, engineering studies, etc.[19] These arrangements played a major role in the transfer of technology to the country, not only because they provided a leverage in bargaining with foreign contractors, but also because the foreign partners had to provide adequate training to permit the replacement of foreign personnel by Algerians.

The exact nature of the experience of the developing countries in these technology transfer agreements depended on a number of factors such as their relations with the international oil industry, the economic and legal environment in which the transfer took place, the level of technological and industrial development, etc. At the risk of over-simplification, some observations of a very general nature may be made.

(a) The contents of technology transfer agreements with developing countries in the oil sector varied widely. There seemed to be more technology transfer through these agreements in the area of production and refining than in exploration. In Mexico's case, patented and non-patented know-how were the most usual components in the agreements involving exploration technology. In almost all such agreements, the training of domestic personnel received important attention.

(b) Parallel to their use of foreign technology, developing countries with fairly advanced domestic oil technology, such as Venezuela, India and Mexico, have begun to make a great deal of effort in domestic research and development in the area of exploration technology.[20] Some of these research and development institutes have started to act not only as domestic sources of oil exploration technology, but also as suppliers in the international market.

(c) Some technological collaboration agreements concluded by developing countries were with 'non-traditional' suppliers of technology from developed countries. India and Iraq, for example, relied quite extensively on technology agreements with the USSR and other Eastern European countries. Several developing countries had technology transfer agreements with developed countries' institutes like the Bureau d'Etudes Industrielles et de Coopération de L'Institut Français du Pétrole (BEICIP), as well as the Japanese and Norwegian petroleum development authorities and private service contractors.

(d) The payments for the technology transferred through these agreements were made by a variety of formulae. In many cases, they were made as lump-sum fees, in others they were made in terms of guaranteed access to oil. In the case of the 1976 Technical Collaboration Agreement of Venezuela with the international oil companies, the arrangements for payments were not related to the nature and value of the technology and service provided, and these were therefore renegotiated and changed in the 1980 contract.

(e) In addition to the direct costs, there appear to be also implicit or hidden costs for the technology transfer, such as transfer pricing and lack of implementation of agreed terms, particularly in cases where the terms were defined too generally.

(f) Finally, the government policy of the developing countries concerned in respect of the development of the domestic oil sector appeared to be a key factor in bringing about some of the new elements in the terms relating to technology in technical collaboration agreements, as was the operation of the subsidiaries of the transnational companies.[21]

CONDITIONS FOR ACCELERATING THE TRANSFER OF TECHNOLOGY

The conditions for accelerating the transfer of oil exploration technology to developing countries may be discussed at two levels — the international level, on the one hand, and the national level in the developing countries, on the other. In the following, some preliminary suggestions are made concerning the issues which may be addressed at each of these levels.[22]

Improvement of the international environment for the transfer

Oil exploration is an expensive and risk-bearing operation. Owing to their strong financial and technological capacity, international oil companies will continue to be involved in exploration and development activities in developing countries. Given this international context for oil development, one way to improve the relative negotiating power of the developing countries, particularly those which are capital-deficient, would seem to be to make finance available from multilateral sources independent of the oil companies. The scope of such a multilateral financial scheme may vary, covering either the exploration or the development phase, or alternatively the geological survey.[23]

Another important way to improve the international environment for the transfer of technology will be to create alternative channels of technology transfer to the hitherto current joint undertakings with the international oil companies. One obvious alternative will be the institution of a multilateral technical assistance scheme for provision of training in all aspects of oil exploration work.[24] Yet another alternative may be to seek expanded co-operation from public oilfield service companies in developed countries.

It may be recalled also that the negotiation of such a code, under the auspices of UNCTAD, is continuing at a United Nations Conference on an International Code of Conduct for the Transfer of Technology. The most recent session of the Conference was held from 23 March to 10 April 1981. Such a code, when established, will provide international norms for the transfer of technology, which will facilitate the acceleration of the transfer of technology under improved terms and conditions.

Initiatives by the developing countries at the national level

There is growing recognition among the oil-producing developing countries of the need for them to develop their domestic technological capacity in the oil sector in order to control and manage their oil resources and to derive maximum benefit from these endowments. To go one step further and institute a technology policy as part of the oil sector development strategy will constitute a critical condition for the acceleration of the transfer of technology. For those developing countries which are initiating oil exploration activities, attention may be focused more sharply on the capacity to negotiate exploration contracts.

In order to link conception of technology policy needs to actually implementable measures, it is necessary to define the type of skills and know-how whose acquisition would constitute the objective of the technology policy. While the exact requirements may vary from country to country, one may generally place the following four categories of skills as critical to technological capacity for planning, negotiation, and implementation of oil sector development:

(a) Managerial expertise for organizing the oil exploration and development programme as a whole.
(b) Technological expertise for taking decisions at critical stages of exploration and development.
(c) Technical capabilities and knowledge of the oilfield service and equipment market for appropriate procurement policy.
(d) Oilfield skills, in particular those of tool pushers and maintenance personnel.

The first two skills (i.e. managerial expertise and the technological expertise for taking critical decisions on exploration activities) are essential for the exploration phase and for the negotiation of contracts. For example, pre-drilling exploration work requires management which could aim for many alternative seismic surveys so as to provide the best information possible while making cross-checks on costing and results and at the same time providing training opportunities for the domestic personnel. At this stage the ability to take key technical decisions is vital, since various geophysical and geological reports require analysis and interpretation. A favourable exploration strategy may seek to time these different activities in such a way as to facilitate training and maximize opportunities for securing experience.

The second two skill categories pertain more to the stage of project implementation. Without procurement know-how and knowledge of the market, the exploration authorities in the developing countries would have to depend on the intermediary firms and also be unable to monitor their pricing, the quality of the materials procured, etc. Furthermore, unless they take procurement policy into their own hands, the developing countries will have to forego an important policy instrument for encouraging domestic suppliers of services and equipment. The importance of the fourth category (oilfield skills) is obvious, particularly for those countries in which the oil sector constitutes an important industrial base.

Acquisition of the four categories of technological capacity mentioned above may be pursued in a number of ways. An appropriate and necessary first step will be the reassessment of the role of oil exploration contracts in the transfer of these skills. The cursory review made in the preceding section indicated that so far both the risk and the service contracts concluded by developing countries have played only a minor role in this respect. Much, on the other hand, will be gained from learning from the experience of certain countries which have negotiated innovative contracts with expanded technological components. Considerable work yet remains to be done in drawing appropriate lessons from such experience.

A second step related to the first is this: in those countries where oil endowment prospects are particularly good, special attention should be paid to the relative merits of service contracts in extending national control over the oil sector activities, which in turn will have an important implication for the domestic learning opportunities. Even short of a complete policy shift, increasing participation by the national oil development authorities in decision-taking on exploration activities — for example, through the establishment of a joint management board — would be a useful step in the same direction.

Third, diversification of the sources of technology will also help accelerate the transfer of technology through a strengthening of the bargaining options of the country concerned. Although the international oil companies have strong technical advantages in exploration and development, particularly for offshore activities, an important part of the technology is today supplied in the international market by independent oil service companies. In addition, the multilateral and public sources in developed countries may also be tapped for certain oil exploration technologies and technical assistance. Possibilities therefore exist, particularly in relation to onshore activities, for the oil-searching developing countries with sufficient capital resources to avoid acting through the intermediary of the international oil companies, provided that they possess the capability to plan and organize the exploration and development.

Fourth, state oil companies in developing countries that have accumulated considerable organizational, managerial and technical know-how in this area can also be an important source for technology transfer.[25] These companies and independent service companies can become partners for developing countries willing to explore their own resources while building up their technological capacity. There are also possibilities for co-operation among developing countries at the regional level, as exemplified by the recently established oilfield service company of the Organization of Arab Petroleum Exporting Countries (OAPEC). Institutional arrangements to bring them together to perform such a role need to be explored.

Finally, new initiatives by developing countries to strengthen the domestic technological capacity in the oil sector will, by necessity, have to be considered in the comprehensive framework of petroleum development strategy. The approach employed in the present meeting to shed light on all aspects of petroleum exploration provides a good example which may be followed by developing countries at the national level within their own developmental context. Such initiatives should also be assessed within the framework of national science and technology policy. UNCTAD, through its Advisory Service on Transfer of Technology, will be ready and willing to act, upon request, in an advisory capacity in this latter respect.

NOTES AND REFERENCES

1. In preparing this note, the UNCTAD secretariat has benefited from a paper prepared for it by Dr T. Turner on the technology components in oil exploration contracts, which is being finalized for publication later on, as well as from discussion with Dr Eleodoro Mayorga-Alba (Energy Division, United Nations Economic Commission for Europe).

2. According to a survey commissioned by the World Bank on the oil and gas situation in 70 developing countries, 23 countries have prospects of finding more than 750 million barrels of petroleum and a further 15 countries of finding between 100 and 750 million barrels. For reference, most African countries consume less than 5 million barrels a year. See World Bank, *A Programme to Accelerate Petroleum Production in the Developing Countries*, January 1979, p. 17.
3. UNCTAD Secretariat, 'Energy supplies for developing countries: issues in transfer and development of technology' (TD/B/C.6/31/Rev.1).
4. See Bain, J.S., *Barriers to New Competition.* Harvard University Press, Cambridge, Mass., 1956.
5. Formerly, the international oil companies provided the technical services themselves. Progressively, however, they came to rely on highly specialized independent contracting service firms. For a detailed discussion of this development, see Añez, C.M., 'International transfer of technology for oil exploration and production, with special reference to the Venezuelan oil industry', Ph.D. dissertation, University of Sussex, 1978 (mimeographed).
6. In addition, there are also numerous autonomous consulting firms which provide a variety of services ranging from prefeasibility and feasibility studies to preparation of project plans and organization and supervision of tender offers. See 'Ownership of petroleum technologies' (ESA/NRET/AC.11/BP/3) by M. Hiegel (Institut Français du Pétrole), background paper presented at the United Nations Symposium on State Petroleum Enterprises in Developing Countries, Vienna, 7-16 March 1978.
7. See Hiegel, M., *op. cit.*
8. See Adelman, M.A., *The World Petroleum Market.* Johns Hopkins University Press, Baltimore, Md., 1972, pp. 34-37.
9. See 'Exploration for natural resources in developing countries: financial requirements and institutional mechanisms' (ESA/NRET/AC.12/1), March 1978, pp. 7-11, paper prepared with the assistance of consultants by the United Nations Centre for Natural Resources, Energy and Transport for the Intergovernmental Group of Experts on Mineral and Energy Exploration, April 1978.
10. See Hossain, K., *Law and Policy in Petroleum Development.* Nichols, New York, 1979, pp. 118-120.
11. During the 1970s, some OPEC member countries concluded service contracts under which they provided all the risk capital for exploration and hired the company and independent contractors for supply of services and equipment. Risk contracts also have variants ranging from traditional concessions to joint ventures and profit-sharing arrangements, but they need not enter into the present discussion of technology components.
12. This survey was made by Dr T. Turner as part of a forthcoming study undertaken for the UNCTAD Secretariat on the technology components in oil exploration contracts.
13. In order to assess the seriousness of this limitation, a sample of contract summaries was compared with the original contracts, and the representation of the key terms in the summaries was found to be accurate.
14. See, for example, Megharief, D. and Turner, T., 'Transfer of oil technology: a case study of the Libyan experience'. New York, February 1980 (mimeo).
15. For example, provision of technical advisers, managers and construction overseers by United States majors in an oil-exporting country in the Middle East involved a fee estimated at between 15 and 21 cents a barrel.

16. See Prasad, N.B., 'Role of the Indian engineering industry in the development of oil resources in the country', *Lok Udyog — Public Enterprise* (New Delhi), Vol. XI, May 1977.
17. See Malhotra, A.K., 'India's offshore oil programme from policy to plans' and 'On the development of indigenous capabilities for offshore production systems'. January 1978 (mimeo).
18. Añez, C.M., *op. cit.* (see note 5 above).
19. A brief account of technology transfer experience of several developing countries in the oil sector as a whole is given in UNCTAD Secretariat document TD/B/C.6/31/Rev.1, *op.cit.* (see note 3 above), paras. 204-248.
20. In India, the Oil and Natural Gas Commission devotes 2-3% of its revenues to in-house research activities. The research establishment there also includes an institute of petroleum exploration, an institute for reservoir studies and an institute of drilling technology. In Mexico, the Mexican Petroleum Institute (IMP) provides an important part of the technological requirements of the domestic oil industry. In Venezuela, an Institute of Oil Technology (INTEVEP) has been created to undertake research and development in this area.
21. See Sherif, S., 'Activities and technical capabilities of state petroleum enterprises' (ESA/NRET/AC.11/3), paper submitted to the United Nations Symposium on State Petroleum Enterprises in Developing Countries (Vienna, 7-16 March 1978). See United Nations Centre for Natural Resources, Energy and Transport, State Petroleum Enterprises in Developing Countries. Pergamon Press, New York, 1980, pp. 48-58.
22. The UNCTAD Secretariat prepared and issued in October 1980 a document entitled 'Major technology issues in the energy sector of developing countries' (TD/B/C.6/65).
23. The World Bank has initiated a special oil exploration facility (*op. cit.*, note 2 above). In the United Nations, the General Assembly, in its resolution 33/194 of 29 January 1979, called for the fielding of United Nations missions to developing countries to assess their financing needs for exploration of natural resources (including primary energy resources).
24. A United Nations intergovernmental group of high-level experts discussed this question as part of the problems concerning multilateral development assistance for natural resources exploration, and, based on its report, work has been initiated for provision of such technical assistance.
25. An interesting example is that of Braspetro, a subsidiary of PETROBRAS of Brazil, working under a service agreement signed in 1972 with the Iraq National Oil Company. Braspetro has already made two major strikes near Basra in southern Iraq; its first success occurred in 1976 when it discovered the Majouroon field, which has a production capacity of 350,000 barrels per day. The Oil and Natural Gas Commission of India has also undertaken exploration activities in Iran and the United Republic of Tanzania.

9

THE FORMULATION AND DESIGN OF APPROPRIATE PETROLEUM EXPLORATION PROGRAMMES

Martin Forrer
Exploration Consultant, New York, USA

This paper is devoted to a discussion of exploration programmes, the reasons behind them, and their appropriateness. Other papers have discussed the technology, the available methods, their costs — others again have dwelt on the problems involved in negotiating or financing exploration programmes. At this point it may be worthwhile to ask ourselves what we are actually doing when we explore. How does one start exploration? Why do we negotiate, and then mobilize all these technologies? How do we know that we are doing the right thing (or have done the right thing in the past)? If a discovery results, then the answer to the last question is 'yes' (even though there may be doubts as to whether the same result might have been obtained faster or more cheaply by different programmes). However, when the results are negative it is often difficult to know whether we carried out the right kind of programme, whether it was adequate. Yet it is now known that the distribution of hydrocarbon deposits is governed by laws which are becoming better understood. Discovery more and more often results from logical search programmes. An appropriate or adequate search programme is of course important for a variety of reasons:

(1) An appropriate programme is more likely to result in economic success.
(2) Future programmes have to be built on the foundation of earlier adequate programmes.
(3) Appropriate programmes avoid wasting money and they are neither over- nor under-designed.

An added consideration is the fact that very often in negotiations between host governments and oil companies, the question of an 'adequate' programme is difficult to resolve. In the absence of clear criteria, recourse is often had to statistical or geographical approaches. That is, areas covered in their entirety on a 1 × 1 km seismic grid are considered more adequately explored than others with less seismic coverage. No one can deny that there is a purely quantitative relation between area and programme. A 100 km seismic line cannot adequately explore a very large basin; on the other hand, is a dense grid *a priori* appropriate? What if the basin has only a few hundred metres of sediments?

The fact is, of course, that hydrocarbons occur in a variety of traps and in diverse geological conditions, and no paper can claim to show how to explore or lay down a final rule for an appropriate programme. Exploration is partly an art; it uses tools and techniques but their results must be analysed and skilfully interpreted and combined by the explorationist. Diagnosing the data, judging their relative value and true meaning, combining known facts and clever ideas into the hypothesis to be tested by an appropriate programme is as much an art as is a doctor's diagnosis leading to further treatment. The application of this art is bolstered by experience and requires, one suspects, an innate talent for handling data and seeing the potential combinations and their meaning. As such, exploration is difficult to describe in words, but some of the philosophy of approaching exploration, and some of the guiding principles, can be stated in an attempt to see what might constitute an appropriate programme.

If we cannot use a statistical approach requiring so many miles for so many square kilometres, what then can we use to judge? Perhaps the answer lies in a quick analysis of why any area gets explored at all. Sometimes, because like Mount Everest, 'it is there', but more often owing to an idea, a hypothesis in some explorationist's mind. The idea or hypothesis is formulated through critical analysis of the data, and through skilful combination of what is known and what may be inferred or stipulated. The idea may not always be explicit to everyone concerned but, make no mistake, it is always implicit, and requires a specific programme to test it. A clearly formulated hypothesis regarding the prospects of an area allows the design of a programme aimed at checking the hypothesis, and avoids both under-designing the programme, or the 'let's-hit-it-with-all-we've-got-and-see-what-falls-out' approach, which in the long run is very wasteful, and certainly not appropriate.

All exploration programmes share one basic problem: how best to test the underlying hypotheses and to arrive ultimately at the point where a well will be drilled as the final test. An 'appropriate programme' might be defined as one that can find all useful deposits in an area in the shortest time and at the lowest cost. That is not a good definition. Firstly, hydrocarbons occur under such a variety of conditions that it is impossible to design from the start a programme to explore for all possibilities in a cost-efficient manner. Secondly, one is rarely, if ever, interested in all deposits at the same time: the largest and/or most economic ones are the early targets. Thus, exploration programmes have more restricted aims than total resource discovery. They try to narrow down the prospects step by step, starting with the most promising.

These remarks may demonstrate a firm grasp of the obvious, but all too often exploration programmes violate this basic principle: that an appropriate programme must test a hypothesis, must search for the answer to a specific question or questions, analyse and proceed to the next step, and do this in the shortest time and with the least expenditure. The question may vary from the very general ('Can hydrocarbons occur in this area at all?') to the very specific ('Exactly where do they occur in this square kilometre and at what depth?'). The questions tend to become more complex with progress, and, of course, programmes may be designed to answer more than one question at the same time. A basic criterion for judging whether an exploration programme is appropriate is to see whether the problem to be solved has been clearly identified and whether the programme addresses this problem in the most efficient and expert manner.

An underlying assumption is that the hypothesis must be worth the testing. It is not in anyone's interest to spend money looking for deposits that are non-economic. Economic considerations may differ, depending on whether they are stated by a government or a transnational corporation, but within their differing framework, certain deposits will clearly be uneconomic. Owing to differing economic ideas, defining an appropriate programme is often a hurdle in agreements between governments and companies. A compromise must usually be found between a government's desire to explore an entire area for all possible deposits, no matter how small, and the transnational's goal of spending money only to look for deposits which are commercial by its standards. To resolve an impasse, a company will often agree to an over-designed programme by simply considering the extra cost as a kind of bonus payment. However, a poorly located early well is of little use to either a government or a company, and it drains funds better spent on a more appropriate project. A better strategy for governments to obtain an appropriate evaluation is to define the minimum programme they require and then to use stringent relinquishment terms to cause the operator to explore rapidly and efficiently.

Thus far, our concern has been with programmes which are either under way now, or planned. What about the appropriateness of past programmes? It is often difficult for governments to judge whether past programmes were adequate, especially when the results were negative. Firstly, one may ask what were the economic assumptions which shaped the past programme. Exploration done 15 years ago to find the larger size fields which were then economical may have been underdesigned for today's oil prices, which make smaller fields profitable. Also, certain fields can now be exploited economically owing to advances in technology, and these opportunities may not have been the object of a thorough search 15 years ago. Another judgement may be based on the tools and techniques used in past programmes. Geophysical methods, for example, have made such progress in the last 15 years that perfectly appropriate programmes of that time are no longer acceptable today. New programmes should be considered.

These are some criteria for judging past programmes. But again, the most important judge is the analytical explorationist. It is he alone who takes the negative results of past exploration, analyses them, sees whether they can be circumvented, and formulates the new hypotheses which justify a new programme. The more extensive the past programme, and the more negative its results, the more important is a creative explorationist. A top-notch geologist will use the positive results, will find excuses for or explain away the negative results (or even turn them into helpful pointers), will not be unduly bound by dogmas, and will discern those prospects, formulate those hypotheses, which justify a new programme. This is an important point, especially for those 80 or so developing countries where little or no exploration has taken place. The attitude of pessimism on the part of the countries themselves (if it exists) should give way in many cases to the realization that geological prospects can be difficult to see and are often obscure, but they can also often be 'created' by good explorationists. They start as a gleam in a geologist's eye, as a concept in his mind. This concept can be tested by an appropriate programme and such programmes are 'promoted' or 'sold' to management, i.e. the financiers (whoever they may be — government itself, oil companies, banks) on the basis of their merit. Developing a worthwhile idea and selling it is the daily bread of an explorationist, and he can do this for a country as easily as for an oil company, once the country has decided to ask him to do it.

An appropriate programme (old or new) may thus be defined as the programme which at any given time in any given area provides the needed answers to a positive hypothesis in the shortest time and at the least cost. The programme design requires the following input:

(1) A first-rate, clever explorationist full of optimism, basic knowledge, and experience.
(2) Analysis of the relevant information already available.
(3) Based on this analysis, a firm understanding of the next hypothesis to be tested.
(4) Up-to-date familiarity with available tools and techniques, and with the practical know-how of their application.
(5) A realistic, but optimistic, economic evaluation of the minimum size of what will be a useful deposit so as not to over- or under-design the programme.
(6) A solid appreciation of the money, and time, available for a programme.

The process of exploration can be divided into four steps, each requiring the shaping of appropriate programmes:

(1) Preliminary evaluation phase. Generally relatively cheap, to decide whether the area is prospective at all, where the prospects may be, how good they could be, and to formulate the hypothesis and programme for the next step.
(2) Primary (pre-drilling) exploration phase. More costly geological and geophysical subsurface investigation.
(3) Drilling phase. Most expensive.
(4) Post-drilling exploration phase. Programme depending on positive or negative results from drilling.

It can be said that governments may be involved in all four, but there are two phases in the above scheme where government involvement is perhaps most easy. First, in the preliminary evaluation, which requires modest investment but can give a good assessment of the value of an area and hence aid the government in negotiations. Second, governments may consider serious involvement again in the post-drilling phase if drilling has been successful and will (soon) begin to generate a cash flow.

As exploration proceeds through the four phases, programmes become generally more complex; it is easier to define an appropriate programme in the first stage than in the last. In the preliminary first phase, the basic hypothesis is often merely this: the right kind of rocks to generate and trap petroleum occur in the area. The question thus is 'Does the area have petroleum potential at all?', and the appropriate programme will search for already available information on the geology. A major input will be a study of the geology of the total region. Its geology and hydrocarbon data can often be extrapolated to judge a specific area. Unless the early search is totally negative, this will allow the designing of an appropriate follow-up programme, which will try to determine the prospective areas more exactly. This may need days or months of surface geological mapping, assisted by aerial photography, side-looking radar, laboratory analysis of rock samples (age, type, porosity, permeability, geochemistry, core drilling), etc. In offshore areas, the appropriate programme is somewhat different since such methods as surface geology, etc., are not really applicable. After an analysis of sea bottom data, topography, drainage, dredging results, etc., and a study of the adjacent land and sea areas, the early offshore programme soon proceeds to geophysical work. Basic questions as to the presence, thickness and structuring of sediments can often be answered quickly and cheaply by a few well placed seismic lines. At this stage, both onshore and offshore, the appropriate programmes may have resulted in a fairly good assessment of those parts of an area that are prospective, and of their possible worth.

These early aspects of the programme may perhaps have been overelaborated, but there is a reason: as a basic exploration strategy, governments should be much more often involved in early exploration than they have been in the past. For instance, too few developing countries with marine areas undertake a small seismic programme to find out if they have any prospects and where, and then use this information to shape their policy on future exploration. Just this basic exploration information can be an invaluable help in shaping a policy of going it alone, or one involving outside expertise and risk capital. The point has been made that a variety of ideas and approaches by different explorationists is an advantage in exploring an area, and one basic strategy question can be answered with much greater confidence after some preliminary exploration, namely, how large an area can and should be assigned to any single operator, and what the minimum work (exploration) programme should appropriately consist of. The area may be large and prospective enough to attract ten operators with ten programmes rather than only one, and the sum of the parts in exploration is generally larger than the total. One should have an idea of the value of an asset before selling (i.e. leasing) it.

The resolving power of these early programmes is, however, limited: oil and gas occur at depth, and this preliminary work is not an infallible guide to conditions below ground. Much better information is needed before proceeding to drill. There are, for example, large swamp- or desert-covered land areas, where even after all preliminary work, there is still no reasonable picture of the distribution of rocks below ground. In a second phase, therefore, geophysics and perhaps early deep drilling are appropriate to give the needed subsurface information, albeit at a much higher cost than anything done so far. In part, the preliminary programme had the objective of identifying certain less prospective areas, so that the more expensive geophysical second phase could be concentrated on more promising areas only.

The second stage proceeds to elucidate the subsurface conditions in the better areas, at first regionally. The value of understanding the regional structure cannot be overemphasized, since oil and gas are so obviously tied to certain regional elements. An appropriate programme must, therefore, consider regional seismic lines on land, or a not too dense grid of seismic lines offshore, or else some indirect field methods, such as gravity or air magnetics. The decision as to which method is appropriate depends on the geological problem to be solved and sometimes on the money available, but also quite often on subjective experience. The aeromagnetic survey is much less expensive than a seismic survey for the same area coverage, but it is also much less definitive in its results. Aeromagnetic surveys are rarely sufficient on their own to eliminate follow-up seismic work. The writer is inclined to by-pass aeromagnetic surveys any time in favour of seismic work unless there are special circumstances. Others feel differently and thus there may be honest differences of opinion as to what is an appropriate programme. To judge, one must have a firm grasp of the basic problem to be solved, e.g. defining regional structure to the depth and confidence level needed to base further exploration decisions on the results. Where the geology allows this to be done with aeromagnetic techniques alone, this method is of course sufficient; where the geology is more complex and/or without a magnetically definable basement, other methods must be chosen.

Indirect field methods must adhere to a fairly rigid pattern of survey lines and cannot easily be designed to test only specific local problems. A seismic survey may be used to cover an entire basin, but it can also be designed to fit a particular problem. Thus, if economic considerations show that only fields with 100 million tons of oil will be economic, then seismic dip lines at a 15 or 20 kilometre spacing are likely to indicate where such fields may lie. Such considerations are particularly appropriate in land areas, with their very high seismic cost per kilometre; in marine areas, surveys can often be run with a denser grid from the start owing to lower costs. Even then, there is an appropriate grid dictated by a geological hypothesis and economics which is often violated; that is, surveys are over-designed simply because they cost less, or because of the attitude that one can never have too much information (no matter whether it is critical or not!), and so as to avoid having two or more seismic campaigns before drilling. Several surveys are, however, almost always a necessity anyway for appropriate structure definition unless the first survey is wildly over-designed. There are several possible appropriate programmes at this point, and striking a balance between over-design on the one hand, and the cost, delay and problems of bringing in a crew several times, on the other, is one of the more thorny problems in appropriate programme design and operation.

As a general rule, after the regional structure is sufficiently defined, an appropriate programme must be aimed at defining drilling sites for the largest and/or most economical prospects. Since the largest known accumulations of oil and gas in the world are in structures, the appropriate programme is a search for, and definition of, economic-sized structures, or traps.

During this second stage, other surveys are occasionally advocated as being 'appropriate'. These are, for example, surface geochemical surveys, radiometric surveys, electric, micro-magnetic and magneto-telluric surveys. This writer found that a seismic survey is always required sooner or later in order to reach the level of confidence for drilling, and cannot see that these other surveys are often appropriate. Some day, there may be a breakthrough, but at this point they seem only to delay getting on with the needed seismic work.

If the early geophysical surveys were successful, there remain now only rather local areas of interest. An appropriate programme must define enough economic-sized traps to justify bringing in one or more drilling rigs. Appropriately defining a trap generally means having enough evidence to show, within reason, that there are no escape routes for the hydrocarbons. This level of confidence is highly subjective because it is largely based on individual experience and varies with innate caution or tolerance to risk. Confidence is generally reached with fewer data on large simple anticlines than in small complex structural or stratigraphic traps. An appropriate programme will be one that checks all parameters critical to closure to the point where one interpretation of the data stands out clearly as the most logical, and a programme is inappropriate if it goes beyond that point and wastes money on non-critical information. For non-technical people faced with judging whether such a programme is adequate there may be some solace in the thought that if the operator is willing to spend his money on drilling, there is at least one 'expert' who thinks the preparation was adequate.

The third, or drilling phase, drills and tests the defined traps. Given equal ease of access, etc., the largest, most promising and best-defined traps are tested first. This first phase of drilling is often followed by a period of analysis and reinterpretation of data and then a second or third drilling campaign. An appropriate programme of evaluation by drilling depends very much on the number of defined traps, their size, and their geological variety, i.e. are the defined traps so similar geologically that a test of one or two will confirm or condemn all the others, or is there a wide variability, all requiring independent testing? An appropriate programme must test prospects in different geological and geographical locations, and must test different reservoirs and sections to different depths. Every area has an 'economic basement', i.e. a level below which no economic production may be expected. An ideally appropriate testing programme has, geophysically or otherwise, gathered all information down to and including this level, and has drilled well selected prospects all the way down to this economic basement. Here is, by the way, another means of judging the appropriateness of older programmes inasmuch as no area can be considered fully explored unless it has been mapped at economic basement level and sample-drilled to the same level.

Exploration drilling programmes also need to be designed in the appropriate mechanical manner, that is, the drilling programme must allow for proper evaluation of the well results by logging, testing, and sampling. Appropriate programmes are complex and require close co-operation between the explorationist and drilling and mud engineers, log analysts, geophysicists, etc.

The fourth or post-drilling phase can have a wide variety of appropriate programmes. If drilling was successful, this will have given a tremendous impetus to further exploration, since there is hope that the venture will become a commercial proposition, and since a discovery proves that the prospect and the basin have 'worked' in terms of petroleum generation and trapping, it is possible now to proceed to other prospects with much more confidence. Even where no oil and gas have been found, a new phase of exploration almost always opens. After analysis of the situation, an appropriate further programme may be aimed at defining other prospects for drilling: the seismic results may be reinterpreted or reprocessed in order to accommodate the information from the wells, and further wells may be drilled close to early wells on better information, or other traps may now be tested. Often, new seismic programmes may be needed, sometimes on very small areas and involving new techniques, such as seismic, etc.

If the information was very negative, the operator may decide that further expenditure cannot be justified. Before abandoning the area, he may farm it out to another operator who still has money and is willing to try his new ideas and his better techniques. Most modern contracts allow for such assignments, and this is a legitimate appropriate programme for exploration. The value of new ideas and new approaches is hard to overestimate, and proper farm-out provisions encourage their application. Many governments may feel that if an operator is no longer willing to bear the total burden of further exploration, he should return the area to them. The chances are, however, that the government will not be able to re-contract this now 'slightly damaged' area to a new operator for better terms than it obtained from the original operator, while in a farm-out the government loses nothing and avoids the problem and delay of finding a new operator.

Often, after the first drilling effort, and sometimes long before that, the operator realizes that his prospects are not housed in structures but in stratigraphic traps. Once such stratigraphic traps are legitimate exploration objectives, appropriate programmes take a somewhat different form and employ other methods. These will include earlier use of seismic surveys, with a locally much denser grid than in a structural search. The programme may have recourse to special seismic and geological techniques, affecting the manner of gathering field data and processing them. Existing well and production data may be submitted to special physical and chemical analysis. Stratigraphic trap search programmes usually require earlier drilling, and more drilling before a trap is precisely located. All this makes the search more difficult and more expensive, which is why this type of programme is usually reserved for particularly favourable areas, and is not appropriate as a rule in remote, hostile or relatively unexplored areas.

In conclusion, there is for every area, at any given time, an appropriate exploration programme, which can be formulated relatively quickly. However, it has to be expertly designed one step at a time, to handle specific hypotheses. It is not possible or desirable to design an exploration programme way ahead of time, and to 'anchor' it, for instance, immutably in a contract. No programme will remain truly appropriate once information starts rolling in, and great flexibility must be built in. This flexibility accrues to the benefit of everyone concerned. Early on, when very little information is available, an appropriate programme can be formulated in an afternoon, so to speak. Later, and in areas with a long history and about which there is much prior information, ample time must be allowed for analysis of past data before formulating the future programme. This may take several weeks, especially if seismic work has to be reprocessed and reinterpreted. The analysis will in no case be costly. An adjunct to these remarks might be that it behoves governments to treat exploration information with proper care so that it may be available at some future date for study.

Appropriate programmes can and must be designed for each exploration effort. In judging whether a programme is appropriate some of the thoughts in this paper may be useful and recourse may be had also to specific expert advice, since no two programmes are ever alike. Something else that should be borne in mind is that there may be legitimate differences of opinion as to what constitutes an appropriate programme, even within one government office or one oil company. United Nations agencies, including the World Bank, other bilateral programmes, private industry, and consultants are available to provide analysis and formulation. These same groups are also ready to assist in supervising ongoing programmes, or to review proposed ones. The cost of such advice is minimal compared to the cost involved in any programme and, above all, to the value of what may be involved for any government. From early policy decisions on strategy, such as whether and how to undertake exploration (through negotiations or go-it-alone state efforts), to the final production of revenue, an appropriate exploration programme can be of assistance. Exploration is a specialized skill, not taught in schools and only acquired through practice, and the good explorationist is a rather rare species. Many developing countries may find themselves without that expertise in house, and may furthermore find the delay and cost of developing that capability on their own very high and hard to justify in the early stages, since one never knows how often or how long these skills will be needed. They are, however, available to anyone making the basic policy decision that something shall be done about exploring the country's own area.

PART III

Financial Aspects of Petroleum Exploration

10

FINANCIAL REQUIREMENTS AND METHODS OF FINANCING PETROLEUM EXPLORATION IN DEVELOPING COUNTRIES

Francisco R. Parra
Director, International Energy Development Corporation, Switzerland

CURRENT LEVELS OF EXPLORATION IN DEVELOPING COUNTRIES

Many people connected with the international petroleum industry would agree today that not enough exploration is taking place in developing countries, particularly in oil-importing ones, and that there is an urgent need to increase it sharply. The definition of 'enough' is debated by geologists and economists (usually among themselves rather than between them, since neither understands the other), but the curious fact remains that exploration in developing countries did not respond noticeably during the 1970s to the large oil-price increases of 1973 and 1974.

Current levels of petroleum exploration in developing countries are not significantly higher than in 1970, prior to the price increases, with seismic party months increasing by only 1.2% per year, and exploratory drilling by only 2.3% per year.

	Exploration in developing countries (excluding OPEC countries)		Increase (% per year)
	1970	1979	
Seismic party-months[a]	1402	1561	1.2
Exploratory wells drilled	970	1194	2.3

[a]Twenty-nine countries only. Comparable data not readily available for other developing countries.

The reason for the poor response to price increases is that there are a number of constraints on petroleum exploration in developing countries which are not directly related to either the geological prospects or the level of oil prices. Some of these constraints concern conditions in the developing countries themselves (e.g. exclusion of foreign companies on policy grounds, inadequate legal framework, general political risk, border disputes, etc.); others, which are of greater importance, pertain to the structure of the petroleum industry and the competing demands of exploratory prospects in the OECD area, especially the United States.

The constraints caused by conditions in the developing countries are becoming less acute, those stemming from the industry's structure and the demands of the OECD area more so. On balance, it seems doubtful at present whether any great upsurge in exploration in developing countries will take place as a result of the second round of large oil-price increases which occurred in 1979 and 1980.

EXPLORATION RESULTS

Although results are very mixed by countries, it is clear that they have, in the aggregate, been relatively meagre. Excluding Mexico, and the major OPEC countries, crude production in developing countries increased at only 4.6% per year from 1970 to 1980, while petroleum demand rose slightly faster, at 4.9% per year. The gap to be filled by imports is thus widening.

	Production of and demand for oil (thousands of bpd) 1970	1980	Increase
Production[a]	2483	3907	1424
Demand[a]	3483	6400[b]	2435
Difference	1482	2493	1011

[a] Excludes major OPEC countries and Mexico; includes other exporters

[b] Partly estimated.

The increase in production has been of the order of 140 thousand bpd each year, while demand has been rising at 240 thousand bpd each year.

About 40% of total production in developing countries (excluding major OPEC countries, Mexico and Argentina) now comes from fields discovered during the years 1971–1978 inclusive.

If we look back over the 1970s, we can see just how meagre the results of exploration in third world countries have been. The average production of the 122 fields discovered in developing countries (excluding Mexico and the major OPEC countries) during the period from 1970 to 1978 inclusive, is only 11.1 thousand bpd. If the few relatively large fields found during that period are excluded, notably the Bombay High, the average drops to only 9.8 thousand bpd. Who is going to search for these needles in the haystack, a search which requires more tenacity and more money than the search for the big fields for the obvious reason that it is easier to find the big ones than the small ones?

THE COST OF EXPLORATION

There are only sketchy data available on historical costs of exploration in developing countries. Geological and geophysical expense rose from $400 million in 1970 to $1085 million in 1978 (13% per year in nominal terms, much lower in real terms). These figures include the OPEC area and Mexico, which account for about half of the total. Given the above figures, the amount of seismic work now being done and the number of exploratory wells being drilled, a plausible guess at current exploration expenditures in developing countries (excluding the OPEC area and Mexico) might be of the order of $3.6 billion per year.

	Billions of dollars
Seismic work	0.3
Exploratory drilling	3.0
Other	0.3
Total	3.6

It is difficult to adduce much evidence for this 'guesstimate'. $1.1 billion was spent on geological and geophysical exploration in developing countries in 1978, but this figure includes the major OPEC countries, Argentina and Mexico, all of which are excluded from the above estimate. Thirty-two United States companies spent $2.5 billion on exploration outside the United States and Canada in 1980, but this figure includes the rest of the OECD area. The national oil companies of Brazil and India will between them be spending almost $1 billion per year on exploration in the next few years, but they will account for a major proportion of all exploration in the developing countries. None of these figures is inconsistent with our own figure of $3.6 billion for an over-all level of recent spending on petroleum exploration in the developing countries specified.

Clearly, much more needs to be done if sufficient discoveries are to be made to close the widening gap. If, in the future, exploration does

Table 1
Developing countries:[a] crude production 1970 and 1980

	Production (thousands of bpd)		Change, 1970—1980	
	1970	1980	Total	Annual
Africa				
Angola	103	163	60	6.0
Cameroon	—	43	43	4.3
Congo	neg.	61	61	6.1
Egypt	326	564	238	23.8
Gabon	109	177	68	6.8
Ghana	—	8	8	0.8
Ivory Coast	—	1	1	0.1
Tunisia	87	110	23	2.3
Zaire	—	20	20	2.0
Total Africa	625	1147	522	52.2
Asia				
Brunei	138	230	92	9.2
Burma	16	37	21	2.1
India	142	203	61	6.1
Malaysia	18	282	264	26.4
Oman	333	287	-46	-4.6
Pakistan	9	10	1	-0.1
Philippines	—	14	14	1.4
Syria	81	161	80	8.0
Turkey	69	45	-24	-2.4
Total Asia	806	1269	463	46.3

Latin America				
Argentina	390	492	102	10.2
Barbados	—	1	1	0.1
Bolivia	24	24	—	—
Brazil	168	182	14	1.4
Chile	33	32	-1	-0.1
Colombia	220	111	-109	-10.9
Ecuador	4	228	224	22.4
Guatemala	—	4	4	0.4
Peru	72	201	129	12.9
Trinidad and Tobago	141	216	75	7.5
Total Latin America, excluding Mexico	1052	1491	439	43.9
Total Africa, Asia and Latin America	2483	3907	1424	142.4
Mexico	487	1989	1502	150.2
Total all developing countries	2970	5896	2926	292.6
Note: Demand	1970	1979	Change, 1970—79	
Africa[b]	493	789	296	32.9
Asia[b]	1457	2694	1237	137.4
Latin America[c]	2015	2899	884	98.2
Total	3965	6382	2417	268.6
Mexico	459	1128	669	74.3
Total	4424	7510	3086	342.9

a Excluding major OPEC countries.
b Same countries as listed above.
c Excluding Mexico.

Table 2
Developing countries:[a] crude production, first half of 1980

	Fields discovered in	Production (thousands of bpd)	No. of fields	Average size of fields (thousands of bpd)
Africa	1970 + before	499	44	11.3
	1971 – 1978	590	42	14.0
	All	1089	86	12.7
Asia	1970 + before	817	63	13.0
	1971 – 1978	418	32	13.1
	All	1235	95	13.0
Latin America[b]	1970 + before	670	97	6.9
	1971 – 1978	345	48	7.2
	All	1015	145	7.0
Total above	1970 + before	1986	204	9.7
	1971 – 1978	1353	122	11.1
	All	3339	326	10.2
Mexico	1970 + before	268	68	3.9
	1971 – 1978	1561	30	52.0
	All	1829	98	18.7
Total all	1970 + before	2254	272	8.3
	1971 – 1978	2914	152	19.2
	All	5168	424	12.2

a Excluding major OPEC countries and Argentina.
b Excluding Argentina.

not result in higher levels of new production per dollar spent, then the total number of dollars spent must rise sharply. Something like a doubling of total exploration expenditures in developing countries is called for if production is, in the aggregate, to keep pace with demand.

FINANCING EXPLORATION: SOURCES

Since early 1979, oil prices have risen from about $13 per barrel to about $35 per barrel. This provides a powerful incentive for increasing exploration in developing countries. If such an increase does in fact occur, how will it be financed?

Private sector

Large oil companies

Turning first to the large oil companies, the so-called 'majors', as well as other large companies, some of which are equally big though more concentrated in the United States, these are the ones that historically have done most of the exploration in the third world. A number of them have been retrenching rather than expanding, and concentrating more and more of their exploration budget in the OECD area, and the United States in particular. Some of them have virtually disappeared from the international scene at the third world level. So it is doubtful whether any great upsurge in exploration will come from the major companies (including in the term 'major' the other large United States oil companies).

European and Japanese state companies

A second group of companies is the European and Japanese state oil companies, included here in the private sector because, viewed from the third world, they do act very much as private sector companies. Some of these are quite active, but it is doubtful whether they will give any great impulse to exploration in developing countries because they are primarily motivated to search for crude oil in prospective countries where exportable surpluses might be developed. Like the majors, they are in it for access to oil, while the need and more likely prospects for many developing countries are to find the small fields yielding sufficient oil to meet their own domestic requirements.

Small independents

There are so many of these that it is difficult to count them or keep track of their activities. At least 80 hold a participation in acreage in oil-importing developing countries, but few of them hold acreage 100% for themselves and few of them hold participations in acreage in more than one or two countries. In the aggregate, their effect is limited and they seem to be even less likely candidates to provide the motive power for an upsurge in exploration in developing countries during the next decade.

It should be noted that all three of these groups, large companies, state oil companies, and independents alike, are engaged in exploration in developing countries only as a relatively small part of their total business. Their main-line business is in the OECD area, and, for most of them, in the United States.

Local private sector companies

In one or two countries there exists a small body of local private sector

Table 3

Developing countries:[a] crude production during the first half of 1980, by year of discovery (in thousands of bpd and numbers of fields)

	Year of discovery									
	1970+ Before	1971	1972	1973	1974	1975	1976	1977	1978	1st half 1980	No. of fields
Africa											
Angola	97 (10)	2 (1)	10 (2)	23 (3)	—	9 (2)	—	—	2 (1)	143	19
Cameroon	—	—	6 (1)	—	11(1)	12 (1)	26 (1)	—	—	55	4
Congo	27 (2)	—	27 (1)	—	—	—	—	—	—	54	3
Egypt	239 (15)	4 (1)	21 (2)	139 (2)	101 (1)	—	8 (2)	9 (1)	48 (2)	569	26
Gabon	43 (11)	18 (3)	—	28 (1)	31 (3)	6 (1)	20 (3)	—	2 (1)	148	23
Ghana	—	—	—	—	—	—	—	2 (1)	—	2	1
Tunisia	91 (5)	4 (1)	1 (1)	—	—	—	—	—	—	96	7
Zaire	2 (1)	—	—	18 (1)	—	—	—	—	2 (1)	22	3
Total Africa	499 (44)	28 (6)	65 (7)	208 (7)	143 (5)	27 (4)	54 (6)	11 (2)	54 (5)	1089	86
Asia											
Brunei	202 (6)	—	—	—	—	21 (1)	—	—	—	223	7
Burma	30 (5)	—	—	—	—	—	—	—	—	30	5
India	85 (18)	—	—	—	100 (1)	—	—	—	—	185	19
Malaysia	92 (6)	29 (2)	73 (2)	27 (1)	4 (1)	29 (1)	26 (1)	—	—	280	14
Oman	197 (4)	—	48 (3)	23 (1)	10 (1)	2 (1)	—	—	—	280	10
Pakistan	12 (5)	—	—	—	—	—	—	—	—	12	5
Philippines	—	—	—	—	—	—	—	14 (1)	—	14	1
Syria	163 (4)	—	—	—	2 (1)	—	—	—	—	165	5
Turkey	36 (15)	3 (2)	—	4 (3)	—	—	1 (2)	1 (6)	1 (1)	46	29
Total Asia	817 (63)	32 (4)	121 (5)	54 (5)	116 (4)	52 (3)	27 (3)	15 (7)	1 (1)	1235	95

Latin America											
Barbados	1 (2)	—	—	—	—	—	—	—	—	1	2
Bolivia	22 (10)	—	—	neg. (1)	—	—	2 (2)	1 (1)	—	25	14
Brazil	128 (20)	5 (2)	1 (1)	9 (1)	6 (1)	17 (2)	13 (1)	—	4 (2)	183	30
Chile	16 (9)	—	—	—	—	—	—	8 (2)	7 (2)	31	13
Colombia	109 (23)	2 (1)	5 (1)	—	1 (1)	—	7 (1)	—	—	124	27
Ecuador	200 (9)	1 (1)	1 (1)	neg. (1)	neg. (1)	—	—	—	—	202	13
Guatemala	—	—	neg. (1)	—	3 (1)	—	—	1 (1)	—	4	3
Peru	70 (8)	18 (1)	5 (3)	90 (4)	neg. (1)	neg. (1)	—	neg. (2)	14 (3)	197	23
Trinidad and Tobago	124 (16)	—	84 (3)	—	40 (1)	—	—	—	—	248	20
Total Latin America excluding Mexico	670 (97)	26 (5)	96 (10)	99 (7)	50 (6)	17 (3)	22 (4)	10 (6)	25 (7)	1015	145
Total above	1986 (204)	86 (15)	282 (22)	361 (19)	309 (15)	96 (10)	103 (13)	36 (15)	80 (13)	3339	326
Mexico	268 (68)	3 (1)	190 (3)	326 (4)	303 (6)	21 (1)	515 (2)	144 (7)	59 (6)	1829	98
Total	2254 (272)	89 (16)	472 (25)	687 (23)	612 (21)	117 (11)	618 (15)	180 (22)	139 (19)	5168	424

a Excludes major OPEC countries, and Argentina, on which insufficient data are available.

companies engaging in exploration. They have limited resources and are of virtually no importance in the world scene, though they may be important in the countries where they operate.

Public sector

National oil companies

While some increase in spending on exploration in developing countries can be expected from the group of companies listed above under Private Sector, these are unlikely, in the writer's view, to provide the kind of impetus needed for a rapid increase in effort, in any of the oil-importing developing countries, but particularly those with only marginal prospects. It therefore seems logical that the national oil companies of developing countries will have to play a much larger role in the future than had been anticipated. In many countries this will pose difficult questions on priorities for government spending, particularly because exploration is a high risk activity, and the national companies have no way of spreading their risks in the way international companies do. Moreover, with limited resources even when accorded priority for exploration by the government, the national companies will want to turn increasingly to international institutions for aid, and these have been notably reluctant in the past to finance the most expensive phase of exploration, namely, wildcat drilling. There are, however, some hopeful signs that attitudes are changing on this score.

It may even be possible to envisage a group of national oil companies from developing countries, joined perhaps by one or two like-minded national companies from industrial countries, setting up an international oil company of their own. This would have the great advantage of spreading risk, as well as affording the geologists and geophysicists a wider experience than they could gain exclusively in their own countries. It is an idea to be treated with great caution, however, because the degree of complementarity necessary for a joint undertaking of this kind may not exist in sufficient degree among the national oil companies of developing countries. Some have a high level of expertise, others might have access to surplus funds, and some have attractive acreage needing money and technology or development, but there are many pitfalls, including, most obviously, political ones in what could otherwise be a concrete and fruitful form of South-South co-operation.

The World Bank

The World Bank's proposed energy affiliate would double spending on energy from about \$12 billion to \$25 billion over a period of several years. The World Bank's activities in the field of energy are the subject of a separate paper at this meeting and, beyond noting that the Bank may be an important source of finance for petroleum exploration, the topic need not be further covered here.

OPEC

The five-man ministerial committee of OPEC, which drew up a long-term strategy plan for approval by the organization, recommended that substantial funds be made available to developing countries on a grant or soft-loan basis to assist them in developing indigenous energy resources and particularly hydrocarbons. It was specifically envisaged that exploration for petroleum should be an important target of such aid. OPEC's present internal difficulties have led to the suspension of the plan for the time being, but there appears to be some talk of its being revived.

Multilateral aid

The principal other example of multilateral aid that affects petroleum exploration in developing countries is the agreement or agreements between Mexico and Venezuela on the one hand, and nine Central American and Caribbean countries on the other. Under these agreements an amount of money equal to about 30% of the purchase price of oil acquired from Mexico and Venezuela is made available by these countries to the Central American Caribbean nations in the form of long-term soft loans. The basic terms are a 4% interest rate and five-year repayment period. However, if the loans are used for capital investment projects in the economy, especially the development of indigenous sources of energy, then the loan terms become an annual interest rate of 2% with a 20-year repayment period.

Bilateral aid

There has so far been very little bilateral aid directed specifically or exclusively at exploration. One of these few examples is the substantial aid made available by the Federal Republic of Germany to Bangladesh, which has provided enough money and technical assistance to do geophysical work in certain areas of the country and drill up to about 10 exploratory wells.

FINANCING EXPLORATION: MEANS

Since exploration is a high risk activity, means of financing tend to be restricted to foreign risk capital, capital from national sources channelled through the national oil company, and as yet relatively marginal aid funds. The present section of this paper will focus on foreign risk capital discussed in the context of the mechanisms, new and old, used to encourage exploration.

Exploration and production agreements

These have a long history encompassing a varied range of possibilities from the traditional concession to joint ventures, production sharing contracts, risk contracts etc., with many variations within each broad type of contract. There is a considerable literature on the subject, dealing primarily with comparative economic evaluation of the various possibilities and with the legal implications for the relationship between state and foreign private investors. As far as the economic side of the agreements is concerned, their development has largely been directed towards ensuring that the host government obtains all or most of the economic rent or windfall profits from large fields, should these be discovered. For a number of historical reasons, this is natural and it is of course necessary for governments to avoid repeating the mistakes of the past and allowing the foreign investor a disproportionately large reward for his efforts in the event of the discovery of a major field. But in fact this is largely a problem of the past. In more recent agreements attention has shifted towards maximizing work obligations on the part of the foreign investor (a trend which is accompanied by de-emphasis of signature bonuses).

It is now necessary to go one step further and develop models which will not only protect the government in the event of large low-cost discoveries, but will also encourage the search for and development of smaller fields, which is the most that many developing countries can hope for in the light of their geology, but which would be extremely important to their balance of payments. Work is being done by several people and institutions on the development of production and exploration agreements specifically designed to permit and encourage what have hitherto been considered as marginal fields. My own company, the International Energy Development Corporation (IEDC), has done considerable work on the development of a new concept which both recognizes the fundamental distinction between the high risk exploratory phase and the relatively low risk development phase of exploration and production and, second, provides for majority participation by the national oil company in the development phase if the government so decides. The main elements of the proposed new concept are:

(1) The foreign company would normally fund the exploration programme and take all the exploratory risks.
(2) In case of success the foreign company would undertake to develop the field on behalf of the host government or its national oil company. The national oil company would normally be the major investor in the development of the field, using funds borrowed from international sources.
(3) The operating costs would be born by the project itself. The foreign company's exploration cost and its share of development costs would be recovered from the first net operating revenues, along with and at the same rate as the host government's recovery of its share of development costs.
(4) After initial exploration and development costs have been recovered, an exploration profit (as distinct from any other type of profit)

would be provided to the foreign company at a level that would recognize the high element of risk involved in exploration.

(5) Following payment of exploration profits there would be an additional, ongoing revenue payment to the foreign company that would constitute a small percentage of net operating revenues for the remainder of the project's life.

(6) Funds loaned to the host government for investment in development would be repayed from the government's share of net operating revenues according to the terms agreed by the lending institution.

(7) Access to crude oil supplies would be negotiated as a totally separate item, independent of the revenue sharing provisions of the agreement.

Financing seismic work

A commonly used mechanism today consists in the government borrowing of funds to finance seismic work, with the results then being sold to companies interested in entering exploration and production agreements. This has proved useful, but of limited application on the whole.

Open area geological and geophysical stage

At present, it is standard practice for governments of developing countries to give exclusive permits for exploration at the geological and geophysical stage. Little consideration has been given to any alternative. However, it is conceivable that governments could open areas to non-exclusive exploration up to the drilling phase, subject only to guarantees that they would obtain all information, including interpretation of seismic work, and that the environment and local private property rights would be protected.

There are a number of potential advantages in such open area exploration. First, it is likely that more exploration in the aggregate would take place and that the government would have access to the results. Second, it would give different interpretations of the same or similar data. Third, it would shorten lead-times significantly, by permitting exploration to start before the negotiation and finalization of an exploration and production agreement. Finally, it does not preclude (although it does delay) the possibility of exclusive exploration rights being granted in the event that no interest is shown by foreign companies in undertaking exploration without the assurance of exclusive development and production rights.

Revolving fund for exploration

Suggestions have been put forward from time to time, including two papers presented at this meeting, whereby governments of foreign companies could borrow from a fund set up specifically to finance exploration, with repayment into the fund being made only if the venture is successful. This suggestion has not prospered so far, partly, perhaps, because of the lack of success of the existing United Nations Revolving Fund for Natural Resources Exploration (excluding hydrocarbons). The present fund is, however, under-financed and there has in addition been an unwillingness on the part of governments to make replenishments to the revolving fund out of the proceeds of successful ventures which would necessarily have to cover the cost of failures elsewhere.

Exploration insurance fund

I was pleased to see that one of the United Nations papers has taken up an idea on which we have done some work and which this writer first presented in outline in a seminar last year jointly sponsored by OPEC, OAPEC and UNCTAD. The exploration insurance fund is fundamentally different from a revolving fund because it would not in itself provide any financing for exploration, but only coverage for failure. In essence, the idea is as follows:

(1) Development assistance funds would be needed to set up the exploration insurance fund.
(2) Its probable size would be between $500 million and $1 billion, spread over a period of 8 to 10 years.
(3) The funds would be available to provide insurance coverage against failure to find oil or gas under a previously approved (by the fund) exploratory programme which would be initially financed either by foreign risk capital or by a national oil company.
(4) Replenishment of the fund would come out of successful finds.

The advantages that the fund presents are considerable. First, financing requirements are relatively small, amounting to about $15 million to $100 million per year of funds committed (but not necessarily disbursed). The first payment would not be required until about three years after the fund is set up, at the earliest. Ultimately, it would be self-sustaining.

Second, with respect to flexibility of coverage: the fund could cover 100% of the total exploration programme, or it could cover something less than 100% of the total exploration programme; or alternatively it could cover 100% of, say, a second phase of the exploration programme.

Third, with respect to flexibility of replenishment: this could be in the form of a fixed fee of a certain number of dollars per barrel, up to a maximum amount to be negotiated; or it could alternatively take the form of a small carried interest assigned to the fund.

Needless to say, the foreign company's rewards would be scaled down to take account of the degree of risk taken by the company. The insurance coverage itself would be provided not to the company but to the host government.

If success is achieved before the fund is exhausted, then replenishment would start, and it might be possible not to have all commitments called.

Finally, and most important of all, is the likelihood that credible insurance coverage of exploration risks would make exploration for the first time a bankable proposition, that is, exploration could be financed on a project basis by commercial banks.

New institutions

There is a need for new institutions whose main-line business would be exploration and development in developing countries. It was noted above that all of the exploration being carried on today by foreign companies is undertaken very largely as a sideline of these companies' main activities in the OECD area. IEDC is perhaps today the only company whose main focus of activity is exploration in the developing countries. It is, in addition, a company without any dominant nationality among its shareholders, which are, in equal parts, Canadian, United States, Swedish and Kuwaiti. Its management is also international in character and the shareholders themselves are partly private sector and partly public sector. It is itself a private sector company and must eventually, in the nature of things, realize a reasonable return on its investment if it is to continue as a viable entity in the field of exploration in developing countries. It is to be hoped that other similar companies concentrating on exploration and production in developing countries will be created.

The key to increased exploration in third world countries does not of course lie with any one mechanism or type of institution. In the end, all alternatives will have to play their part foreign oil companies, national oil companies, development assistance funds, new mechanisms for financing; none is exclusive of any other, and the extent of the effort required is large enough for there to be more than enough room for all.

11

THE WORLD BANK PROGRAMME TO ACCELERATE EXPLORATION IN DEVELOPING COUNTRIES

Philippe Bourcier

Assistant Director, Energy Department, World Bank, Washington, DC, USA

INTRODUCTION

In 1970 the oil import bill for the developing world was \$5 billion; by 1975 it had increased six-fold; and today it is \$50 billion — ten times its 1970 level. The change has been traumatic; in many countries the cost of oil imports has risen from insignificant levels to account now for over 50% of export earnings. Yet per capita energy consumption in the developing world is still relatively low; growth in their energy use is an inevitable and essential part of their development process. To achieve even modest growth objectives total energy consumption in the oil-importing countries will have to double over the next decade. Under any conditions imported oil will remain an important part of total energy consumption in these countries. In the absence of a vigorous effort to expand domestic production of energy, the oil import bill could rise to a staggering \$111 billion by 1990.

The World Bank's present and projected energy programme was developed against the background of these grim statistics. Building on important existing programmes for electric power and coal, the Bank now lends for all forms of primary energy, including renewables. Further, efficient energy use has become a major concern in preparing all loans to energy-intensive sectors. The Bank's comprehensive approach to third world energy problems is described in *Energy in the Developing Countries*, World Bank, August 1980.

Of these relatively recent initiatives, the petroleum lending programme is perhaps the most important, and it has grown rapidly in response to the escalating urgency of the need to contain oil import costs. Assistance to petroleum development began in 1977. In January of 1979 an Accelerated Programme was approved which included financing for exploration or pre-development activities. Since mid-1977 the Bank has committed over $1 billion for 25 projects in 20 countries. Half of these projects are primarily for support of pre-development activities, and many of the projects designed to support development activities will have a positive impact on exploration as well. An additional $1.3 billion in Bank financing for 40 petroleum projects is due for approval before the end of the 1982 fiscal year. Nearly three-quarters of these projects in preparation are pre-development projects.

The need to do something to correct the plight of the oil-importing developing nations is surely clear, but there has been some debate over which actions are appropriate. Before elaborating on the content of the Bank's programme to accelerate exploration, I should like to spend a few minutes on its rationale, on the question of its appropriateness.

PROGRAMME RATIONALE

The first premise of the Bank's programme, or indeed of any expanded exploration effort in the oil-importing developing countries, is that a reasonable potential for increased domestic petroleum supplies exists. Based on an analysis of individual country prospects, the Bank believes that a major exploration effort could result in additional oil output of 1.5 to 2 million barrels per day by 1990, a 75% increase over 1980 production from the same area. Exploration for gas, which has been neglected in developing countries except where there is a potential for export, could add another 1 million barrels per day of oil equivalent to domestic energy supplies by 1990.

If the potential for new production is there, is the Bank needed to realize it? The international oil industry has operated successfully without the Bank in the past and one might not unreasonably ask what special circumstances now exist to explain the Bank's presence. This question brings us to the second premise of the Bank's programme: the existence of a number of major obstacles to the involvement of private capital and expertise in oil-importing developing countries. These obstacles or constraints include:

(1) Host country inexperience in promoting exploration activity.
(2) Political risk perceptions.
(3) Fundamental differences in the viewpoints of host countries and oil companies. For example, because of taxes, royalties, etc. the economics of additional exploration will always look better to the host country than to the oil company. And even if this difference were not important, the urgency of a developing country's need for additional investment will typically be greater than the priority assigned to such investments by private oil company investors.
(4) Lack of domestic infrastructure.

(5) Inexperienced or inadequately financed local partners.
(6) Inappropriate government policies in such areas as incentives to foreign investment and energy pricing.

Obviously, these constraints do not all apply in any one country at one time, but one or more exist in varying degrees in most developing countries. Problems in countries with established production are more likely to be of a policy nature. Where production is not established, inexperience and absence of local institutions or infrastructure are obstacles of paramount importance.

Sometimes these obstacles may be overcome by time. For example, recent trends show an increasingly reasonable and pragmatic approach to foreign investment policy among host countries. However, under present circumstances, time, or delay, is the most expensive of all solutions. The Bank's petroleum programme is designed to speed up the dismantling of these obstacles; it is intended to act as a catalyst to increase and accelerate the flow of capital to the petroleum-dependent third world.

Complementarity and not competition is the key to the Bank's relationships with other sources of funds and expertise. Even aside from the substantial merits of this approach, the numbers involved would make it inevitable. The Bank's current programme for petroleum exploration and development probably represents no more than 6 - 8% of the oil-importing third world's upstream petroleum investment requirements over the same time period. The annual sums allocated to exploration projects alone would amount to less than the annual exploration budget of one medium sized oil company, and be spread over many more countries. The Bank energy staff has visited 50 countries over the past two years and identified over 40 exploration projects. Obviously if the Bank is to be effective, the nature of its participation in a country's petroleum sector, i.e. programme content, will be crucial.

Having identified the rationale of the Bank's programme, I should like to spend my remaining allotted time reviewing the actual content of the programme: how does the Bank play its chosen catalyst role? How does it go about reducing barriers to the flow of capital and expertise?

PROGRAMME CONTENT

The Bank's exploration activities fall under either of two broad headings: direct action or indirect action. Direct actions include projects to promote exploration interest; reduce investor perceptions of political risk; and, in specific circumstances, finance exploration or appraisal drilling. Indirect actions include local infrastructure financing; institution building; and assistance in the development of policy and planning. Let us look at each of these in turn, starting with the direct actions.

Direct actions

Exploration promotion

Exploration promotion projects are designed to bring a country quickly to the point of offering new acreage to the international industry on reasonable terms. Projects typically involve one or more of the following: preparation of a promotional data package; technical assistance and training in exploration matters; expert advice on petroleum laws and contracts; and assistance in negotiations. Exploration promotion loans range between \$5 million and \$8 million and activity occurs over 6 - 12 months.

The most costly element is preparation of the data package, which may involve not only collation of existing data, often in disarray and gained over many years by different contractors, but also reprocessing of data and limited acquisition of new reconnaissance seismic data. In some cases we expect that the government will recover a substantial part of its costs through sale of the information obtained to oil companies.

While less costly than preparation of a data package, expert advice on laws and contracts and experienced assistance during negotiations may be of equal or greater importance. The presence of experts familiar with industry standards and practice can benefit both host governments and companies in terms of easier negotiations and improved prospects for contract stability. Advice to host governments on technical exploration matters contributes in a similar manner to an improved negotiating atmosphere. I should make it clear that in any exploration promotion project the Bank does not take any advocacy position; rather it makes itself available to governments and investors alike in an 'honest broker' capacity.

Through its exploration promotion loans or credits the Bank is effectively financing the 'set-up' costs of a country exploration programme. In most countries targeted for Bank attention, private capital to undertake the same investment is not forthcoming either because the expected petroleum potential does not justify an individual company's allocation of the time, money and manpower to come up with an evaluation or because initiatives by interested private parties are likely to prove ineffective (e.g. in the area of advice on laws and contracts).

Even where limited private capital is available, the Bank's programme has some inherent advantages over an immediate and *ad hoc* award of the exploration contracts. These include: a wider ultimate audience of interested companies; accumulation of data on a consistent and regional or country-wide basis rather than on a contract basis; a consistent approach to contract negotiation; and the establishment of a framework within which the host government can design and monitor its exploration policies, and retain the initiative, instead of waiting for proposals to be made by the oil companies.

The demand for exploration promotion projects has been particularly strong. They account for 7 out of 13 pre-development projects now under way and 20 new exploration promotion projects are under preparation for approval before the end of the 1982 fiscal year. In terms of geographic coverage most of these projects are in Africa and the Caribbean. Exploration promotion projects now nearing completion (e.g. in Madagascar, Mali, Liberia, Equatorial Guinea) have already produced substantial benefits for country and companies alike. Starting without any institutional framework for petroleum exploration and without any significant expression of company interest, countries have in a matter of months found themselves with contract negotiations under way or even with new foreign company exploration well commitments. And companies have seen 'back-burner' exploration plays move to possible project status in times which are well under the delay that would be normally expected.

Political risk

There can be little doubt that there are countries where, despite favourable geological potential, exploration is at a standstill because investors judge the political risks too great to commit funds. There are a number of cases where governments have nationalized assets in the past but subsequently have come to realize that private sector participation is essential to early development of their energy resources. Not surprisingly, investors aware of past actions hesitate to get involved or will do so only with some limit on their exposure. The Bank's extensive experience in dealing with developing country governments and its ongoing portfolio of loans in these countries put it in a unique position to play a potentially valuable role in mitigating political risk.

Two Bank instruments have been developed to encourage private investment where perceptions of political risk are important. The first is a 'letter of co-operation' provided by the Bank to the host country in question expressing a willingness on the part of the Bank to consider financing the host country's contractual share of development expenditures in the event of a commercial discovery. The letter of co-operation is accompanied by a contractual commitment on the part of the host government to apply for Bank financing.

The second instrument is the guaranteed loan under which the Bank lends funds during the exploration phase to the locally incorporated subsidiary of an international oil company with a full commercial and technical guarantee from the parent company. Political risk, narrowly defined, is accepted by the host country in a parallel guarantee arrangement with the Bank. Guaranteed loans will cover 20 - 25% of the contractor's minimum obligations during the exploration phase; the cost of the loans will normally be no greater than the company's cost of borrowing in financial markets.

Other approaches to political risk are under consideration in the Bank. Each new situation seems to require or suggest new solutions. There are some common elements in the several Bank approaches, however. For example, the Bank does not, indeed cannot, act as a substitute for conventional political risk insurance in protecting assets. Nor will the Bank undertake to guarantee absolute contract stability; neither the Bank nor any other agency, to my knowledge, has been prepared to insure against what is termed 'creeping nationalization'. What the Bank does seek in this area is a framework, which it is hoped the Bank 'presence' will provide, for continued operations and long-term co-operation between host governments and companies. Principally for this reason, the Bank will involve itself only at the *joint request* of government and company, and only after it has been convinced that contract terms between the government and the company are reasonable for both parties in the light of industry practice and any special circumstances.

To date the Bank has issued two letters of co-operation in Pakistan, and a third is pending. These letters facilitated the negotiation of new contracts with substantial exploration work commitments. A guaranteed loan is now under consideration in an African country in the context of new contract negotiations. Both the country and the company involved have expressed strong interest in the loan; with the guaranteed loan a sizeable exploration programme in a new area will get under way.

Letters of co-operation and guaranteed loans providing for Bank 'presence' are under discussion with respect to a number of other Bank projects. Interest in such formulas is expected to grow as the Bank expands its relationships with smaller or mid-sized independent oil companies. These companies, which have the potential to make a considerable contribution to exploration in developing countries, are typically less familiar with international operations and less willing to accept political risks, real or perceived, than the big oil companies.

Exploration and appraisal drilling

The Bank has identified three possible lending situations involving detailed seismic work, and exploration or appraisal drilling by the host country government. These are: financing the government's share of exploration expenditures where the host government has a contractual obligation to participate with the foreign company during the exploration phase; financing follow-up by the government of exploration 'leads' or marginal discoveries either on acreage relinquished by the foreign oil companies or on acreage under contract through the exercise of 'sole risk' provisions; and financing the ongoing exploration programme of a national oil company where its intended programme is not of interest to, or simply not available to, foreign oil companies. Loan amounts for these activities will vary considerably among projects but are expected to average \$10–\$20 million for the first two situations and up to \$50 million for the third.

When the Bank involves itself in one of these situations it is usually because private capital is not forthcoming or is forthcoming only on the incentive conditions made possible by the Bank loan (e.g. where the host government's share of exploration expenditures is financed by the Bank rather than 'carried' by the foreign investor). Despite the lack of private investor interest, a project not infrequently may be attractive from the host country's perspective. For one thing, even the most generous fiscal terms used in the petroleum industry today will leave a sizeable gap between foreign company and host country economics. This gap may justify country investment where private investment would not make sense. The divergence in host country and oil company perspectives is further increased by the general interest consequences which host countries associate with a successful project, e.g. benefit to the balance of payments, reduced energy dependence, renewed industry-wide interest in an area. . . . These consequences, of great importance to the host country, are of no direct relevance to an individual oil company.

Where host country exploration can be justified on country economics or on general interest grounds, the Bank will still condition a loan or credit on the competence of the national oil company and on some signs that the country's over-all exploration strategy is moving in a reasonable direction, especially in regard to attracting outside capital and expertise. Priority will go to those projects which are likely ultimately to increase private sector interest and involvement.

The Bank's approach to host country exploration is clearly reflected in the Bank-assisted national oil company programmes now under way in Tanzania, Morocco, Bolivia and Egypt.

Turning from direct to indirect actions to accelerate exploration, these fall, as I indicated earlier, under three headings.

Indirect actions

Infrastructure support

Exploration and development interest in a number of countries has been lessened by the lack of essential local infrastructure. This is particularly true when it comes to gas. Preliminary Bank estimates indicate that about 30% of Bank funds allocated to oil and gas over the next several years will be used to finance gas development projects or, more particularly, gas projects directed at meeting the internal requirements of developing countries. The international petroleum industry has to date shown little enthusiasm for acting alone to explore for or develop natural gas where its commercialization would require sizeable and risky investments in processing facilities, distribution systems and local market development. Bank willingness to consider financing investments in essential infrastructure has upgraded natural gas as an exploration and development target.

Thailand provides a good example. Bank financing of 15% of the infrastructure required to utilize gas reserves discovered offshore in the Gulf of Thailand supplied the missing ingredient to agreement between the government and the Union Oil Company on the development of those reserves. Both the government and Union Oil have expressed satisfaction with the way things have worked out, and there

Table 1
Bank petroleum projects approved, 1977 — present (US $ million)

Region	No. of projects	Project cost	Loan amount	Type of project	Countries
East Africa	3	54.8	48.5	2 exploration promotion 1 appraisal drilling (national oil company)	Madagascar Somalia Tanzania
West Africa	3	17.6	16.8	exploration promotion	Guinea-Bissau Congo Liberia
Europe, Middle East and North Africa	9	717.1	360.5	2 exploration promotion 1 exploratory drilling (national oil company) 2 pre-development engineering 1 enhanced recovery 3 gas development	Egypt Turkey Morocco Yemen Arab Republic Tunisia
Latin America and Caribbean	4	145.8	78.5	1 exploration promotion 2 pre-development engineering 1 secondary recovery	Argentina Bolivia Honduras Peru
Asia and Pacific	6	2116.7	776.8	3 oil development 3 gas development	India Pakistan Thailand Bangladesh
Total	25	3052.0	1281.1	7 exploration promotion 2 exploratory/appraisal drilling 4 pre-development engineering 5 oil development 6 gas development	

have been beneficial side-effects as well. The availability of the Bank-financed pipeline has increased private company interest in the development of previously marginal gas reserves on adjacent acreage and in continued exploration in this gas-prone area.

A large part of the proposed Bank lending for oil development resembles the gas lending programme. Bank financed investments in oil distribution and, in some cases, refining facilities, is expected to result in the commercial enhancement of a potential or known resource base, stimulating increased exploration and development activity.

Institution building

Bank participation in the financing of exploration or development activity in any host country is usually paralleled by assistance to government agencies or ministries dealing with petroleum and to the national oil company, where it exists, to increase its organizational, financial and technical strengths. This assistance has been well received by host countries and private companies alike since it better equips the host country, not only to carry out its own programme but also eventually to act as an equal partner with foreign oil companies.

Policy planning

The Bank's approach to energy financing is based on a country's overall needs and options. As a part of this integrated approach the Bank has embarked on a programme of 'energy assessments' designed quickly to identify key policy issues and propose responses to such issues. Projects of this sort may have direct relevance to exploration where questions such as the pace of licensing, or the division of responsibilities between private and public, or foreign and domestic investors are addressed, or of indirect relevance, as in matters relating to petroleum pricing.

PROGRAMME SUMMARY

In conclusion let me go over some relevant figures again.

The Bank has already committed \$1.3 billion to 25 petroleum projects in 20 countries. Forty new projects are under preparation, ranging over 40 countries, with a forecast commitment of another \$1.2 billion of Bank funds by June 1982. About one-half of the programme approved at present is specifically directed at exploration; over three-quarters of the programme in preparation is for exploration. And, as I mentioned earlier, many of the petroleum development projects will have a positive impact on exploration activity.

Even this aggressive programme is not, in our view, sufficient, and an expanded programme is now under discussion. There is a lot still to be done and our programme is still in a state of evolution. In this context we would welcome your comments, criticisms, and project proposals!

PART IV

Exploration and Development Agreements

12

NEGOTIATION OBJECTIVES IN PETROLEUM EXPLORATION AND DEVELOPMENT: THE PRIVATE SECTOR VIEW

Alfred O. Munk
Manager, Foreign Affairs, Standard Oil Company, Indiana, USA

As a representative of the private oil sector, I am here to tell you that if the geological prospectivity of any given country is favourable, as a risk-taking commercial organization we are willing, and in fact have been willing, to work anywhere in the world, regardless of social system, political ideology or cultural environment. All we ask of the host country in return is that our commercial relationship be marked by four ongoing principles: fairness, mutual benefit, reasonable political stability and contractual consistency.

Having opened with this categorical statement, which, I hope, will set the tone for my brief presentation, let me now back up and say how pleased I am to be at this important meeting, to share ideas with many old friends, and to engage in a mutually rewarding dialogue designed to develop new concepts and understanding.

In a sophisticated gathering of this kind there is no need to dwell on the importance of the topic before us. Our very presence, representative of some 100 nations from every part of the globe, representing as many social, economic, and political systems as exist on earth, underlines the urgency of our joint mission, which is nothing less than to find new and imaginative ways to discover, produce, refine and market as much of the world's reserves of crude oil and natural gas as our collective knowledge and skills will enable us to do.

In this enterprise, it is my view that the private sector transnational oil companies have a particular role to play, a role which draws on our long involvement in exploration and production and on the expertise, technology and capital resources which have been the fruits of nearly 100 years of experience and hard work in every corner of the globe.

On the international side, we have learned, sometimes the hard way, that our ability to be successful in oil and gas exploration and production is directly commensurate with our ability to deal with the host country realistically, free of illusions and aware that we are mutually embarked on a long-term, high-risk commercial enterprise whose success depends on the four principles mentioned earlier: fairness, mutual benefit, political stability, and contractual consistency.

Long experience in the oil business has taught us that failure to negotiate or implement a commercial agreement often derives from misunderstanding and misconceptions. It avails little to assess blame when an agreement breaks down. Just as it takes two to tango, so in this case it takes two to tangle, and both parties may frequently bear some of the blame for their inability to work harmoniously together. What matters in the first place is to avoid the pitfalls by a clear, forthright declaration of objectives by both parties and a sensible determination of how those objectives can best be meshed.

At this meeting you have heard set forth, ably and even eloquently, the requirements of the developing world in the area of hydrocarbon exploration. Hard-headed commercial realism dictates that, because of the high risk involved, the private company putting up the capital, taking the risk, bringing men, money and material to the project also has requirements, objectives, and ideas on the negotiation process. Nothing else makes sense. Negotiation, by definition, is a two-way street.

As it is incumbent on the private company to appraise the needs of the host country intelligently, so the host country should clearly understand the mission and objective of a private company and the economic environment in which it works. Let me state one proposition very simply. To operate effectively — in fact to exist — a transnational oil company must have access to the huge capital markets of the West and Japan. To have access to these capital markets the company must have credit, and that credit derives from its ability to make a reasonable return on its investment. In this connection, I should point out that it is often in a developing country's own interest for the oil companies to maintain a good credit rating, since the companies must often guarantee loans in countries perceived to be 'too risky' by commercial banks.

My first point is that it is mutually advantageous if the host country clearly understands the nature of the private transnational company and where the private sector fits into the world economic scheme. The host country can then work with the company in structuring a deal which meets both their needs within a realistic, mutually rewarding framework. For our part, I can assure you that the private sector is willing to structure a very wide variety of deals so long as they meet two mutual needs: (1) for the company, a reasonable return on capital and assurance of access to some crude oil supply; (2) for the host country, assurance that its internal energy needs are met if sufficient crude oil or gas is discovered, and a sense that the company will provide effective training for the local labour force consistent with host countries' development plans. This assurance argues that the company remains a good corporate citizen.

Being a good citizen is important because, as you here well know, the search for oil is not a short-term proposition. Among other attributes, practitioners of the oil business must be patient, and those entering into contract do so in full knowledge that, like marriage, the agreement is for a long period during which there will be ups and downs and considerable compromise and give and take. It is therefore imperative that the host country be happy with its corporate guest. Realistic recognition of mutual objectives at the outset is the best insurance against premature separations and an unwanted divorce. This is the first key point I would like to make.

Let me make a second point. There is also the necessity of recognizing that the oil industry has changed dramatically in the last decade, which is another way of saying that the sum of money available for capital investment has in large measure been siphoned off from the traditional capital markets of the West and is now distributed among several groups other than the companies; OPEC, state oil companies, organizations like the World Bank, and various national and international development agencies have enormous capital bases in 1981 which they lacked in 1970. Whereas the transnational oil companies still occupy an important place in the constellation of world oil actors, we no longer enjoy the exclusivity of past decades, particularly in the upstream exploration and production side, which is of course the part of the business of greatest interest to you. Parenthetically, may I add that my own company, Standard Oil Company (Indiana), being a relatively late-comer to the international scene, has never enjoyed any degree of exclusivity.

What does this mean to the developing nations and to the private companies? We come back to structuring a deal. The facts are that the oil reserves that remain — those as yet undiscovered — are the tough ones to get at, often in remote areas, at great depth, onshore or offshore, in geological formations and physical environments which challenge the most advanced technology. The risks, and the costs, are therefore far greater than ever before. For example, a typical offshore well may cost roughly $10 million to drill. Failure is becoming prohibitively expensive and can only be cushioned by the proportionately few successes which (and I return to my key point) pay a reasonable return on the company's over-all investment. Host countries which recognize that reality and can tailor their deals accordingly are those most likely to find technically well-qualified private sector partners.

A simple equation may suffice to make this point clear. As risks become greater, and the cost of failure even higher, the relatively few successes must somehow pay for the predictable many failures. The implication for the companies is that they must spread the risk and be involved in several areas simultaneously in the hope that one success will compensate for the predictable eight or nine failures.

It is axiomatic that new conditions demand imaginative new solutions. Forced by circumstances and global need to go after the increasingly tougher geological plays, private companies must spread risk by maximizing opportunities, i.e. by investing in a number of high-risk areas simultaneously, counting on one winner to pay for several losers. Host countries should structure contracts which take this new factor into account.

There is a third dimension to the future of oil exploration in the developing world. It is important, we think, for host countries to recognize that the so-called energy crisis and its attendant problems, identifiable to at least the end of this century, are stretching available technological equipment and trained personnel to the limit. It will be no surprise to you that graduates in petroleum engineering and related technical fields are being grabbed up as fast as they graduate from their schools. Competition for good people is fierce. Gaining access to drilling rigs is at a premium. In 1980, for example, the United States had over 3000 rigs in simultaneous operation; this represents an all-time high for the industry. The number of advanced offshore rigs of, say, the semi-submersible variety is too small for the magnitude of the job required. I say all this to indicate that the developing countries' search for oil comes at a time when demand for available people, skills, and equipment far surpasses our ability to meet it. The developing countries are therefore in competition for that relatively small pool of existing expertise. They are also, it must be said, in competition with new areas of exploration in the developed world, including, for instance, Canada, the Mediterranean, the North Sea and Australia. Most transnational oil companies have available to them at any given time more opportunities than they can possibly avail themselves of. I keep coming back to my point: those nations which structure mutually rewarding deals which best meet the conditions of the two parties are most likely to win the competition and attract the relatively few companies with the know-how to define the job and get it done.

A fourth point: it is my view, and here one should speak frankly, that all too often a developing country will let political ideology get in the way of sound economic judgement. More specifically, it has been my observation that in seeking advice on how to set up a petroleum infrastructure, a ministry, the laws that govern it and the terms it will offer, the developing country goes for help to oil-producing nations which it considers politically friendly — and this is perfectly natural — but the circumstances of these politically friendly countries may be, and in fact often are, at variance with its own needs. Those countries which have already reached the stage of large proven reserves may unintentionally mislead those just entering the field into seeking terms which are unrealistic. The reward, in short, must be commensurate with the risk. It is our view that the needs of developing countries are better served in the oil sector by consultation either with private, commercial, non-governmental groups, or with select supra-governmental organizations who have in-house expertise on energy matters and whose only stake in the outcome is the finding and producing of crude oil on the best terms available to the host country, using the most efficient and productive partner available. It should be noted that some private groups are eager to help and advise developing countries in these matters with the intent of establishing good will and with the hope of being considered favourably when future opportunities arise.

As noted earlier, the structure of the oil business has changed dramatically in the past ten years. As the private oil companies have been forced to adapt to new rules, so must the other players, not least the non-oil-producing developing countries. I have tried in this connection to indicate some areas which may give you food for thought. Let me add a fifth idea.

In thinking about your future and the role of oil in it, if it is your desire to bring in a private transnational oil company, consider very carefully the implications of your tax laws. The tax legislation of the host country should be such that it precludes the paying of double taxation by American companies. There is no reason that such legislation should present the host country with any economic penalty whatsoever. More importantly, such a policy would generate substantial advantage for the host country by significantly enlarging the number of foreign companies interested in competing on exploration and development projects. Tax laws in the host country which eliminate double taxation will enable American companies to compete fairly and freely and thus offer the host government the strength of their particular technological and financial resources.

I realize that it is neither easy nor palatable for a sovereign government to modify its tax laws simply to conform to what many of us think are misguided United States tax regulations; however, such action will not reduce the host country take, which American companies can abide. Admittedly, the matter of transnational taxation is complicated, and the host country, in our view, when preparing or amending its tax code, should seriously consider consulting with a private American tax adviser for expert interpretation of the United States tax laws.

American companies, for reasons having to do with the United States tax laws, have traditionally worked at a competitive disadvantage. It can only be to the developing countries' benefit to eliminate that disadvantage, make American companies more competitive, and thereby bring more technically qualified players into the game.

In their search for hydrocarbons in developing areas, private companies can bring not only capital and expertise, but also jobs, and, perhaps more importantly, training. I raise this as a sixth point. I said earlier that among our more pressing problems was lack of trained personnel in the numbers we require. Therefore, we share with host governments the necessity of creating local talent. Typically, after the start-up of producing operations, the private oil company uses very few Americans, Europeans, or other expatriates, thus creating a local job and training base in the host country. We actively seek out and train young engineers, technicians, accountants and other personnel to become involved in our operations. Having said that, let me make the point that in many instances, particularly in joint ventures, host countries insist on companies paying their technical trainees at government salary levels, i.e. at wages lower than those in the private sector. This often results in trainees leaving after a few years on the job and means that, in the long run, we are unable to train people adequately for middle and top management levels. We would make the plea, therefore, that we be allowed to pay the wages necessary to attract and, more importantly, to keep the best young local brains in the all-important energy sector of the developing countries.

Where practical and commercially feasible, we also use local contractors and materials, creating whole new industries in the process. Apart from any other consideration, we save substantially on transportation costs if we can find locally the materials necessary to do the job. There is, however, one caveat: it can be counterproductive for the developing countries to mandate exclusive use of local supplies and suppliers, which can lead to a position tantamount to monopoly. Competition is the touchstone of good workmanship and fair pricing. Some of us have been burned in this area, and I should like to be forthright in saying so.

However, I do not wish to be misunderstood. I am not here to argue that the oil business in the long run is labour-intensive; it is not and never will be. Our operations normally do not represent a solution for countries burdened by large, unemployed populations. Our mission is different, but what I am saying is that, in the accomplishing of it, we can do much in creating service industries which broaden and diversify the industrial and commercial base of a country. Several instances can be cited in Africa and Asia, not to say the Middle East and even Western Europe, where that has happened.

CONCLUSION

It seems to me that the aide-memoire prepared by the United Nations for this meeting is clear and its message urgent. It says in effect that some 60 sovereign nations — most of them developing countries — have not adequately explored for hydrocarbons and cites as factors the lack of adequate financial, technological and personnel capacities. We believe this is well and truly said. The aide-memoire goes on to make a critical observation, and I quote 'the provision for risk capital for exploration remains a crucial obstacle'. My charge here today has been to speak of private risk capital and our objectives in petroleum exploration and development. I submit to you that risk capital is available in the late twentieth century as it has been since the dawn of the industrial revolution. The problem now, as throughout history, is how to attract it.

I have suggested four principles and six avenues worthy of our mutual examination. One last but critical point: it must be forthrightly said that working with a private transnational oil company, with its diversity of stockholders and its dependence on private capital markets, is a very different business from dealing with state oil companies. Both have their place. Both, we assume, are with us for the foreseeable future. The question for you to decide is which best meets your special needs and particular interests.

The private transnational oil company is in business to take risks and is committed, when conditions are auspicious, to gambling very large amounts of capital in exploration activities. In exchange, as I said earlier, we look for a reasonable return on investment. With few exceptions, the state oil company, as a public enterprise, is in a very different position. Operating with public funds derived from taxes and subject to parliamentary or state control, it often faces inherent problems in allocating and investing in very high-risk activities. For perfectly understandable reasons, these are constraints on state oil companies which do not apply in the private sector. Given such constraints, state companies may well find their capital resources more productively employed in the development portion of the business, leaving the high-risk exploration end to the professional risk-takers whose shareholders have knowingly offered their funds for this purpose.

Needless to say, it is my submission and strong contention that the system which built the mighty engine of the United States economy is capable of reaching out beyond its own borders to invigorate and stimulate other economies even though our social, political and even economic systems may be quite different. For a fair return on investment, private risk capital is willing, as it always has been, to take a chance. I will go further: it is the life blood of the free enterprise system to take risks: not wild, irrational leaps but carefully thought out, measured and judicious risks.

If, as the United Nations aide memoire states, 'a significant proportion of the world's petroleum resources which remain to be discovered are in the developing world', then, we say, let us go find them. If it is within the province of science to put a man on the moon and explore with computers the reaches of Saturn, surely there should be no place on earth where our ever-advancing technology cannot explore for hydrocarbons to make life on this planet better for all of us. The private transnational corporations have the will and resources to do the job. In this enterprise we extend the hand of partnership to you, the nations of the developing world.

13

NEGOTIATION STRATEGIES AND OBJECTIVES IN PETROLEUM EXPLORATION AND DEVELOPMENT AGREEMENTS: THE HOST GOVERNMENTS

Michael Tanzer
President, Tanzer Economic Associates Inc., New York, USA

INTRODUCTION

The aim of this paper is to discuss host government negotiation strategies and objectives in the oil exploration and development area within a broad framework — that of seeking economic development with maximum possible economic and political independence. Within this framework, it is assumed that the goal of developing countries is to increase their share of the benefits from, and extent of control over, their oil production sectors.

Given the 'uneven development' in the third world, different host governments may be negotiating various forms of contractual agreements. These range from a typical concession agreement, under which the government is essentially a passive collector of a share of oil profits via royalty and income tax payments, all the way to a 'true' service contract, under which the government takes the risk of exploration, contracts for all necessary services for fixed fees, and receives all the profits. Historically the host government position in the developing countries has tended to evolve from the former to the latter, while passing through intermediate forms such as joint ventures and production sharing contracts. Nevertheless, all of these forms can still be found today.[1]

Since most countries have not reached the true service contract stage, the focus in this paper is on negotiating some form of 'risk contract' with an international oil company. The commonality in all such contracts is that the company supplies the risk capital for exploration and in exchange gets some share of the profits or the oil, rather than a fixed fee. Nevertheless, in evaluating the cost and benefits of any particular form of risk contract, it is useful to consider the alternative option of undertaking oil exploration and development through service contracts. Only in this way can a government then estimate what is the real cost, both in money terms and in loss of operational control, of not being willing to risk its own capital for oil exploration.

OBJECTIVES OF PETROLEUM EXPLORATION AND DEVELOPMENT

It is relatively easy to list the main objectives of host governments in the oil exploration and development area.[2] These would include maximizing the government's: share of the profits from oil production; control over the disposition of any oil produced (both to ensure adequate domestic supplies and to block shipment to undesired places or countries); control over the pace and character of oil exploration, development and production; internal processing of crude oil (both to increase indigenous value-added and to spur industrialization); acquisition of skills and technical training (in order to develop indigenous capability to monitor and ultimately operate the industry); power to protect the human and material environment; and saving of scarce resources, particularly foreign exchange.

Two things may be said in general about this list. First, normally a true service contract would better fulfil all of these goals except the last one; indeed, the risk of scarce capital is essentially the price paid for the additional benefits from a true service contract. Second, there may be conflicts among these goals requiring trade-offs. For example, in negotiating a risk contract a government which wishes to maximize its money profits may have to give up control of a large amount of crude oil. Within a service contract approach, a conflict may exist between rapid development of the oil industry and preservation of the environment. Thus, in working out its negotiating strategies, the host government has to weigh the relative importance of each of its goals.

CRUCIAL BACKGROUND CONSIDERATIONS: THE PRESENT INTERNATIONAL OIL SITUATION

In order for the host government to develop effective negotiating strategies for meeting its goals, it has to begin with a clear understanding of the present and likely future international oil situation. The most crucial overriding aspect of this situation is the potentially greatly increased bargaining power of developing countries with possible oil reserves. This increase results from the post-1973 leap in oil prices and the new perception of oil as a scarce commodity. These have made oil exploration both an enormously profitable and a necessary undertaking for many more parties, private and governmental.

Moreover, as a result of the increasingly successful efforts by OPEC countries to take control of their own oil resources, even the largest international oil companies have been forced to extend their exploration efforts around the world. While much of this effort has gone and will continue to go into 'safe' industrial country areas, such as North America and the North Sea, opportunities there appear relatively limited. Hence, the companies need to reach out to the vast areas of oil-importing developing countries which hitherto had been relatively unexplored because of the availability of high profit crude oil in the OPEC countries.[3]

Evidence of the heightened drive to find new oil supplies is seen from the increasing activities of industrial country governments and financial institutions. Most prominently, the World Bank has recently established a multi-billion dollar loan programme for oil exploration and development for non-OPEC developing countries.[4] This is a reversal of the Bank's historic policy of refusing to give loans to third world governments for the oil sector, on the grounds that ample resources for such investment were available from private oil companies. Since the historic policy helped to shield the international oil companies against competition from governments in developing countries, and thus served the Bank's primary mandate 'to promote private foreign investment,'[5] its abandonment is a striking indication of the enormous pressure from industrial country governments for more widespread oil exploration.

A similar piece of evidence is the recently changed policy of the United States Government's Overseas Private Investment Corporation (OPIC), allowing insurance of investments by United States oil companies in oil exploration and development in third world countries.[6] This reversal of a previous policy of avoiding insurance in the oil and minerals area (owing partly to the heavy losses on insurance of United States copper investment in Chile) was expressly designed to stimulate new supplies.

A final piece of evidence is the increased efforts of industrial country state oil companies to find oil in developing countries. Since the basic goal of these companies is to find assured supplies of crude oil for their home countries rather than to make money profits, they are potentially important new sources of risk capital. The pressing need for new supplies of crude oil is dramatized by a recent Japanese Government decision to provide funds for overseas exploration without any requirement that oil found be supplied to Japan, on the theory that increasing the world oil supply will give Japan a better chance of obtaining its requirements.

While the new policies of the industrial countries both indicate increased demand for oil exploration and provide a new potential supply source, developing countries must be aware that they may involve conditions which limit the host government's policy options. For example, the World Bank's programme generally assumes that a third world government will operate in combination with major international oil companies, rather than carrying out exploration and development itself through service contracts. For example, the People's Democratic Republic of Yemen was recently granted a $9 million credit for seismic surveys, but the Bank states publicly its goal that this would be followed by oil exploration concessions being offered to oil companies.[7]

Again, the new OPIC approach to encourage United States oil exploration abroad also contains a toughened stance by the United States Government towards third world governments. Specifically, the 'nationalization' which is insured against is no longer viewed simply as a complete takeover by a foreign government, but as 'any material unilateral change by the government in the percentage split'.[8] Moreover, it is clearly indicated that any attempt by a host government later to rectify unfair terms of contracts signed now with United States oil companies covered by such insurance would bring down on it the full force of the United States Government and the international financial community.

Thus, in general, while there are positive signs in the new policies of the industrial countries, they make it even more necessary for third world countries to consider carefully all possible oil strategies *before* negotiating and signing contracts which will be binding for many years to come. As the saying goes, 'Act in haste, repent at leisure'.

In recent years some third world countries have been able to capitalize on the improved bargaining situation, develop effective strategies, and negotiate highly favourable agreements. Many countries, however, have not been able to do so, and indeed some contracts have slipped back to the low levels of the 1950s, when third world countries had little bargaining power. Consideration of some of the reasons for this failure may be helpful in reducing some important knowledge barriers which impede the development of effective strategies and better agreements.

A continuing barrier to greater third world power in oil exploration and development has been the historic myths about these activities — myths widely disseminated by the largest international oil companies (particularly the 'Big Seven'), their home governments (the United States and the United Kingdom) and closely allied international institutions.[9] These myths served to monopolize third world oil exploration and development for the big companies, and thus to deter host governments from either entering the industry on their own or seeking a much wider range of companies with which to contract. In summary, these myths are that in these sectors only the big international oil companies have the necessary *technology* and *large amounts of capital* and can *afford the risk of failure.*

As to technology, the facts are that most oil exploration efforts are not carried out by the big international oil companies but by specialized firms which sell their services to them or to anyone else (including governments) for a fixed fee rather than a share of production or profits.[10] As for the required large amounts of capital, this myth is based on a failure to distinguish between the exploration and development stages, and to recognize that only a small part of the total capital needed (perhaps 10% or less) is for the truly risky exploration stage.[11] For the much larger amounts required for development capital, since oil is such a valuable and profitable commodity, any party, government or private, which is successful in exploration can easily obtain loans on favourable terms. Finally, as to the affordability of oil exploration risk, this depends not only on the exploration cost but also on the possible benefits, and on what the cost/benefit ratio is for alternative uses of the country's capital. Given the enormous value of crude oil today, it may well be that the country cannot afford not to take the risk of exploration, if the alternative is to give up control of a large share of its non-renewable oil resources to a foreign company. For example, if a company gets a production sharing agreement with a government on a 75 : 25 basis, the company may risk a relatively small amount for exploration, like $25 million, and if it finds a moderate size field, with 250 million barrel reserves, which at today's prices are worth about $10 000 million, the company will end up with profits of close to 100 times its risk investment!

Finally, the preceding assessment of the current international oil situation indicates that it should be highly favourable to the bargaining power of developing countries. It is particularly important for host governments to understand this because, in many countries which are experiencing very great foreign exchange pressures from oil imports, there is almost a sense of desperation in seeking to find indigenous oil resources as quickly as possible. While understandable, this could lead to acceptance of inferior contracts in order to proceed with oil exploration. A realistic picture of a country's bargaining potential can be a helpful antidote to such harmful tendencies.

ASSESSING THE RANGE OF POSSIBILITIES FOR OIL EXPLORATION AND DEVELOPMENT AGREEMENTS

As noted earlier, conceptually there are two basic alternatives for exploration and development strategies, and the resulting agreements. The first alternative, which we have characterized as a 'true' service contract, is where the government supplies the necessary capital (either using its own funds or through loans or grants) and the necessary technology (either from within or by hiring specialized foreign firms for a fixed fee) and takes the risk of failure, but gets all of the profits from success. The second alternative, which is some form of 'risk' contract, involves foreign companies taking the risk of failure, in exchange for which, if successful, they get some share of the oil and profits.

These alternatives are not mutually exclusive. For example, Vietnam has pursued a dual strategy of saving the most favourable prospecting areas for development by the Government under service contracts while awarding risk contracts for less favourable areas (see below).[12] Indeed, the very fact that a government is known to be considering a true service contract could strengthen its bargaining position with companies that might want risk contracts, since at a minimum it would show that the government is knowledgeable about the oil industry.

The true service contract approach has a long history. It starts with the post-1917 Soviet development of its oil production sector, and carries through events in Mexico after nationalization of the oil industry in 1938 and in China after 1949.[13] More recent examples, which reflect the post-1973 changes in the oil industry, are those of India and Vietnam. In India, in 1974, as part of a dual strategy, for one prospective area the Government purchased a Japanese offshore drilling rig and hired for a fixed fee a United States drilling company to operate the rig and at the same time to train Indian nationals; for an exploration risk of about $20 million, the Government found oil worth more than 500 times as much, all of which belongs to the country.[14] Similarly, in Vietnam after 1975, the Government reserved the most favourable prospects for itself, originally to be explored with technology, capital and training from Norway; these are now to be supplied by the Soviet Union.[15]

Little can be said in the abstract about strategies and objectives for negotiating service contracts. Obviously, one necessity is to analyse thoroughly the market for the various services so as to be fully aware of the range of companies (and countries) which can provide such services. Another important aim is, where service contractors are paid on a 'cost-plus' basis, to develop methods of auditing and controlling the 'costs' of such service contractors. Finally, since most countries using service contracts seek long-run control over the oil industry, there is a need to negotiate procedures for maximum transfer of technology to the host country.

Considerably more of a general nature can be said about strategies and objectives in bargaining for a better risk contract. One overriding need is to focus on the essence of proposed contracts and look beyond their labels — whether they be called 'concessions' or 'joint ventures' or 'production sharing' or so-called 'service contracts' (where the latter are really a form of risk contract in that the contractor effectively gets a percentage of the profits rather than a flat fee for its services). From a narrow economic perspective, the critical points are what proportion of the costs, money profits and crude oil will be allocated to the companies and to the government respectively, and how these proportions will vary over time. From a broader political economy perspective an additional critical concern is the extent of the government's control over the pace and character of development of its oil industry.

More specifically, in seeking the best possible risk contract for an exploration area, two things have to be considered. First, what are the general goals of the various companies that might undertake the risk contract? Second, for each company, what are its irreducible minimum requirements, and how could the host government maximize its own position while meeting the company's minimum requirements?

To address the first question, companies which might undertake oil exploration and development under risk contracts fall roughly into three groups: those primarily interested in money profits (especially non-integrated oil producing companies); those primarily interested in obtaining assured long-term crude oil supplies (particularly the state oil companies of Western Europe and Japan); and those interested in both (virtually all of the integrated international oil companies). In general, since the crude oil seekers are willing to undertake exploration on terms which would leave a higher share of the money profits to the host country, they are particularly attractive exploration candidates for third world countries which are seeking oil primarily as a source of capital rather than for indigenous consumption. Another generalization is that companies which have exploration agreements in many countries tend to be less willing to agree to favourable contracts because of fear of a 'domino' effect on contracts which are less favourable to these countries.

Nevertheless, it should not be assumed in advance that one kind of company is always better to deal with than another, nor that very favourable contracts can be obtained from any company without great effort on the part of the host government. Indeed, the very wide range of today's risk contracts, as measured for example by the share of profits going to the host governments, is only partly explainable by varying geological prospects. Equally critical, if not more so, is the knowledge and bargaining ability of the host government.

By way of example, at one extreme we have risk contracts where the host government gets as little as 50 - 55% of the profits from any oil found, while at the other extreme we have a few risk contracts where the government would get 95 - 97% of the profits. The most striking of the latter are in Vietnam.[16] The key to the high government share in the Vietnamese contracts, which were made with the crude oil seeking state oil companies of Italy and the Federal Republic of Germany, is that, instead of the successful companies receiving outright a share of any crude oil production, they get the *right to buy* a share of such production, at a small discount off world market prices. (In these contracts, up to 45% of any oil found can be bought at a discount of 7 - 10% below market price, which would net the Government of Vietnam about 95 - 97% of the oil profits.) Moreover, these favourable contracts cannot simply be explained by Vietnam's good oil prospects since, as noted above, the best prospects were reserved for service contracts; in fact, actual exploration under the risk contracts has not met with any success.

Finally, in order for a government to make rational evaluations and comparisons of the costs and benefits of risk or service contracts, it is important to have a systematic and consistent methodology. Such an approach is discussed in the next section.

ANALYSING OIL EXPLORATION AND DEVELOPMENT CONTRACTS

A logical starting point for analysing oil exploration and development is to estimate what the government's profit rate on the investment would be if it undertook the risk of exploration on a service contract basis. The first steps in such a process would be as follows:

(1) Obtain as much technical information as possible about the prospect. One source would be data from companies which have previously explored in the area, while another would be undertaking pre-drilling studies, which can be obtained relatively cheaply (particularly for offshore areas).
(2) Estimate from these technical data several scenarios of the quantities of oil which might be found, as well as the associated probabilities.
(3) For each size of field, estimate the associated annual exploration, development, and operating costs necessary to find the field and bring it into production, as well as the normal production pattern over the life of the field.
(4) For each size of field, estimate future streams of gross annual income by 'guesstimating' the future price of oil and multiplying it by estimated quantities produced.
(5) Finally, for each size of field, estimate the annual net cash flow to the government by subtracting from gross income outlays for exploration, development and operating costs.

The next step is to derive from these annual net cash flows an estimate of the government's profit rate on its investments.[17] (One cannot simply divide the total profits by the total investment because this fails to take into account the 'time value of money'.) In our view, the most useful measurement is the discounted cash flow (DCF) profit rate, in part because it is also widely used by the oil companies to analyse their own investment possibilities.

In essence, the DCF profit rate is calculated from the net cash flows by finding an implicit interest rate such that, if the money used for the exploration and development investment had been put in a bank instead (at that same interest rate), the amount of money accumulated in the bank at the end of the life of the oil field would be equal to the total amount of money accumulated from the actual oil investment (assuming that the annual net cash flows from oil production had also been reinvested each year at that same interest rate). For example, if the calculated DCF profit rate from an oil exploration and development investment is 30% per year, this would imply that the 'opportunity cost' to the government of not undertaking the investment would be 30% per year. Put another way, it would imply that if the government can obtain or borrow capital at a cost of less than 30% the oil investment will be worth while.

Finally, before the government's over-all DCF profit rate on its investment can be calculated, account has to be taken of the risk factors mentioned previously. The simplest way to take risk into account is to generate from each of the scenarios a single composite net cash flow for each year, based on the different probability factors. Formally, this would be done by taking, for each year, the net cash flow from each scenario and multiplying it by the probability of the scenario occurring, and adding these products together.

For those who are not familiar with these mathematical techniques of building up net cash flows, calculating DCF profit rates and using probabilistic risk factors, the reasons for their use may appear obscure and the application extremely difficult.[18] In fact, however, from our experience, the rationale behind these techniques, as well as effective use of them, can be readily learned in a short time period. It is crucial to host governments to master these techniques, partly because they allow a systematic approach to the analysis of oil exploration and development and provide powerful tools for integrating the many variables involved. Additionally, since these techniques are also used by oil companies, an understanding of them is a virtual prerequisite for negotiating favourable risk contracts.

Thus, after a government has estimated the likely scenarios and profit rates on its own investment via service contracts, it can then consider the alternative of seeking a risk contract with an oil company. The basic analytical approach would be to evaluate the contractual terms offered by different companies, by superimposing these terms on the different scenarios. From this, the government can calculate how each company proposes to divide the previously estimated annual net cash flows between itself and the government, respectively, as well as the profit rate the company would obtain. From the government's viewpoint, the aim of negotiations would be to push each company to its lowest acceptable profit rate, while structuring a contract which would give the government the maximum possible benefits. A clear understanding of how each provision of a contract may affect the company's calculated profit on its investment is vital to achieving this aim.

For example, the two most critical contractual factors affecting a company's expected profit rate would be its share of production and profits, and the speed with which its investment costs would be recovered. Since speeding up recovery of investment costs increases a company's DCF profit rate (and also reduces its perceived political risk by allowing it to repatriate its capital more quickly), the government can offer this in exchange for giving the company a lower share of profits. Calculation of the precise impact on company profit rates of such trade-offs can permit the structuring of contracts under which, for the same company profit rate, the government can maximize its share of profits over the life of the field. In fact, because oil companies normally seek profit rates far higher than the cost of capital to host companies, the discounted value of revenues in later years is relatively small for the companies. This allows the possibility of making agreements under which a company would get little or no profits in later years, and still be able to achieve its minimum profit rate.[19]

The preceding discussion has necessarily oversimplified the negotiating problem in a number of ways. One is that since it is difficult to find out what is a company's minimum acceptable profit rate, it is not easy to know exactly how far it can be pushed. A second is that for companies which are primarily interested in obtaining assured supplies of crude oil, use of estimated market prices for oil will underestimate the real profit rate to such a company on its investment; in effect, the real value of oil to such companies is greater than is measured by its market price. One way for the government to seek to overcome these (and other) problems is to draft its own model contract (or contracts) and ask the companies to submit, as a basis for negotiations, their proposals on the key features of such contracts, e.g. extent and speed of work programmes, profit shares, rate of recovery of investment, shares of production to be obtained outright or bought at a discount from the market price, etc.

If the government also makes clear its own general priorities as regards these features, and that it will be considering bids from many companies, several important things can be accomplished. First, it will be an announcement to the oil world that the government is knowledgeable about its bargaining power, and this should help to raise the general level of the proposals made by the companies. Second, it will allow individual companies to shape a proposed contract to meet their particular needs, which could then allow them to offer the government more of what it desires (e.g. a company desiring the maximum amounts of any crude oil found might offer the government a maximum share of money profits). Finally, from such proposals and subsequent negotiations the government can obtain invaluable practical knowledge to supplement its more theoretical prior estimates of key assumptions such as minimum profit rates acceptable to companies.

CONCLUSIONS

The underlying assumption of this paper has been that host governments are seeking maximum benefits from, and control over, their oil sector, in order to develop it within the framework of an over-all economic development strategy. In our view, in the long run, such goals can only be achieved by building the maximum indigenous human and material capability to operate all levels of the industry. Since most developing countries at present lack that capacity, a realistic short-run goal, which can serve for a transitional period, is to develop the knowledge and skills at least to monitor and control the oil industry so that the maximum possible benefits can be obtained for the country. In order to achieve this goal, it is absolutely crucial for host governments to allow the relatively short time necessary to gain technical and economic knowledge — knowledge which is a prerequisite for developing effective oil strategies and negotiating favourable agreements.

While we recognize the enormous pressures on host governments to negotiate agreements 'today', we believe yielding to such pressures would be disastrous for the countries. Ironically, it would not even solve the immediate problems since it normally takes at least four to five years from the signing of an exploration and development agreement before the government begins to get any oil or revenues. More importantly, however, governments have a duty not only to the present generation but to future generations also. And, since oil is a precious non-renewable resource, it would be fundamentally irresponsible for any government to negotiate oil exploration and development agreements without having first built a strong capacity to negotiate such agreements and effectively to utilize the benefits from any oil production; otherwise the people of the country would be better off if the oil stayed in the ground until such time as this capacity could be achieved. It is our hope that this paper and the work of this meeting will be a useful step towards helping host governments build that vitally needed capacity.

NOTES AND REFERENCES

1. See Tanzer, M., 'Oil Exploration Strategies for Developing Countries', *Natural Resources Forum*, July 1978, pp. 319-320.
2. For discussion of some of these factors see Hossain, K., *Law and Policy in Petroleum Development: Changing Relations between Transnationals and Governments.* Nichols, New York, 1979, pp. 43-57.
3. For a discussion on the underexploration of the third world see the World Bank report, *Energy in the Developing Countries.* World Bank, Washington, DC, August 1980, pp. 13-17.
4. For the details on the World Bank programme see *ibid*, pp. 10-11, 67-71.
5. For a more detailed discussion of the World Bank's role see Tanzer, M., *The Political Economy of International Oil and the Underdeveloped Countries.* Beacon Press, Boston, 1969, pp. 90-95.
6. OPIC policy from Middleton, C.H. (Director for Minerals and Energy), 'Overseas Private Investment Corporation Energy Program', paper presented to an OPIC seminar on minerals and energy investment in the developing nations, New York, 17 January 1978. This OPIC policy is discussed within a broader context in Tanzer, M., *The Race for Resources: Continuing Struggles Over Minerals and Fuels.* Monthly Review Press, New York, 1980, pp. 70-82.
7. *Platt's Oilgram News*, 1 July 1980, p. 3.
8. Quote from Middleton, C.H., 'Overseas Private Investment Corporation Energy Program', *op. cit.*, p. 8.
9. For a detailed treatment of the 'historic myths' of the oil industry see Tanzer, M., 'Oil Exploration Strategies for Developing Countries', *op. cit.*, pp. 320-321. An additional myth sometimes put forth, regardless of the facts (or lack of them), is that the country has very 'poor' oil prospects, and any company willing to explore is virtually doing the country a favour!
10. See Hiegel, M., 'Ownership of petroleum technologies' (ESA/NRET/AC.11/BP/3), paper submitted to the United Nations Symposium on State Petroleum Enterprises in Developing Countries (Vienna, 7-16 March 1978). Mimeo.
11. See, for example, the experience in India, discussed later in the text, where exploration was about 4% of total costs.
12. For Vietnam's dual strategy see Tanzer, M., 'Oil Exploration Strategies for

Developing Countries', *op. cit.*, pp. 323-324.

13. For a more detailed discussion of Soviet, Mexican and Chinese oil developments, see Tanzer, M., *The Political Economy of International Oil and the Underdeveloped Countries, op. cit.*, pp. 78-79, 227-287 and 288-303, and Tanzer, M., *The Energy Crisis.* Monthly Review Press, New York, 1974, pp. 92-103.
14. Details on Indian service contract from Tanzer, M., 'Oil Exploration Strategies for Developing Countries', *op. cit.*, pp. 323-324.
15. *Ibid.*, p. 324.
16. As to the key question of cost recovery, under the contract with Deminex of the Federal Republic of Germany, while the company pays all costs, only operating and development costs are recoverable, in kind; this recovery takes place over a 20-year period, during which unrecovered costs accrue interest at 9% per year. For the riskier area explored by Agip of Italy, if successful the company is permitted to recover all costs; exploration and operating costs are recoverable over a 5-year period (but with no interest accruing on unrecovered costs), while development costs are recoverable over 10 years, with interest of 9% per year on unrecovered costs. (See *Platt's Oilgram News*, 26 April 1978, p. 2.)
17. The government can also compute its 'social profit rate' on its investment by adjusting the net cash flows to take into account effects not measured by market prices, e.g. that the value to the government of any oil found may be greater than its market price because it furthers industrialization or political independence.
18. A useful illustration of how such net cash flows are developed is presented in Hossain, K., *op. cit.*, appendix B, pp. 225-238. The mechanics of calculating DCF profit rates from net cash flows is presented in Tanzer, M., 'Investment Requirements and Financing of Petroleum Refineries in Developing Countries', *Proceedings of the United Nations Interregional Seminar on Petroleum Refining in Developing Countries* (New Delhi, 22 January-3 February 1973), Vol. I, pp. 223-237.
19. For example, under risk contracts awarded by the Indian Government in 1974 to independent US oil companies, after 11 years of production the Government could take over the whole operation and get 100% of the oil and profits. See *Petroleum Concession Handbook: Supplement 21.* Barrows Company, New York, 1974, p. 33.

PART V

Strengthening National Capacities in Petroleum Exploration

14

THE ROLE OF STATE PETROLEUM ENTERPRISES IN EXPLORATION: EXPORTING COUNTRIES

Abdul Aziz Showail
General Manager, Oil and Gas Division, PETROMIN, Saudi Arabia

It gives me great pleasure to address this distinguished gathering on the role of state petroleum enterprises in exploration. The subject is of vital importance, not only for Saudi Arabia, but for the entire oil industry and the international community.

Given the composition of this gathering, I would venture to say that I need not expatiate on the reasons for the growing interest in petroleum exploration worldwide. The reasons are self-evident. However, the expanded role of state petroleum enterprises in exploration in the oil-exporting developing countries needs some discussion. This is because of its significance in the restructuring of the world's oil economy and the transition from the traditional system to an era in which producing states' effective participation in petroleum activities will increase. Better understanding of this phenomenon could promote very productive trends in the field of petroleum exploration, which constitutes a crucial sector of the world's economy.

In Saudi Arabia, as in many other developing countries, the creation of the General Petroleum and Mineral Organization (Petromin) in the 1960s was the fulfilment of a fundamental act of sovereignty. The Kingdom made the decision to establish for its citizens a state petroleum and minerals organization which could take effective control of the exploitation and development of its precious natural resources in line with the Kingdom's national and international objectives.

I consider it worthwhile to place the history of exploration in the Kingdom within the broad framework of the OPEC countries' experience. According to most petroleum economists, about 79% of world proven oil reserves and 61% of natural gas reserves outside the centrally planned economies are located in OPEC member countries. Within OPEC, Saudi Arabia occupies a pre-eminent position, as its proven oil reserves are estimated to be about 168 billion barrels, approximately 30% of the proven oil reserves in the world outside the centrally planned economies.

What is of crucial importance for a developing country is not only the size of its petroleum reserves, but also the trend in the relationship between discovery and production of this valuable but non-replaceable asset. In this respect, what has happened in the Middle East during the last decade provides a very good reason for the need of increased state activity in all sectors and particularly in the area of exploration.

In the 1950s, 100 billion barrels were added to Middle East oil reserves against a production of 12 billion barrels. During the 1960s, the figures were 180 billion and 33 billion respectively. In the period between 1971 and 1978, however, a reversal took place, and 28 billion barrels were added to reserves while the region produced 60 billion barrels. It is important to notice that a net depletion in petroleum reserves has taken place at a time when the oil-exporting developing countries have started to take effective control over their petroleum resources.

Admittedly, at the same time, some of the newly emerging state petroleum enterprises possessed very little expertise, whether technical or managerial, and in some cases even lacked the financial ability to engage in a technically complex and financially risky field like exploration. Considering this and several other factors, the result was that the Middle East region of OPEC accounted for only 1% of the world's exploratory efforts. In terms of wells drilled in each 10 000 km^2, the density for the Middle East region was 5 wells only, as against 450 in North America, 85 in the Soviet Union and 100 on the average in the world as a whole. Even assuming a much less rosy prospect for oil discoveries in the future compared with the period up to 1970, this distribution of drilling activity worldwide seems to be far from the optimum if maximization of new oil discoveries is regarded as the objective.

I must emphasize that Saudi Arabia proved to be an exception in this case of declining reserves. Fortunately, our proven oil reserves have increased every year since oil was discovered. Despite high rates of production, the Kingdom has not suffered a net depletion of its reserves. Nonetheless, the emerging trends in the new discoveries, namely, small fields, difficult operating and producing conditions, etc., have warned us that it may not be long before such events take place in Saudi Arabia also.

So the question of how we should preserve an adequate reserve of a depletable resource for the Kingdom's rapidly increasing internal needs and vast international requirements has become of paramount significance to us. We have studied several options, ranging from extending the fields' productive life by reducing production, to maintaining an aggressive exploration programme, and including a number of combinations thereof. We have come to the conclusion that, whereas no one single measure could provide the optimal answer, a programme that includes the following four basic elements could prove to be quite adequate.

(1) Immediate stoppage of energy wastage and encouragement of its productive use. We have done this in Saudi Arabia through our huge master gas system, which is the largest of its kind in the world.
(2) Acceleration of the development of renewable and alternative sources of energy. We are participating in several projects related to the development of solar energy, which has vast potential in the Kingdom. We also have a keen interest in pursuing nuclear energy development.
(3) Intensified exploration activities.
(4) Development of a capable and efficient manpower resource base.

With these arguments behind us, I should now like to describe the historical evolution of petroleum exploration in our country.

Saudi Arabia entered the petroleum world picture in March 1938 when, after four frustrating years of exploration, Saudi Arabian and American oilmen produced the first oil well, called Dammam No. 7. After more than 42 years and a total oil production of 29 million barrels, this discovery well is still in production and is an excellent example of the productivity of the Kingdom's oil fields. Yet it was not until after World War II that production in Saudi Arabia started to increase rapidly. Averaging less than 20 000 bpd before 1944, production was up to 500 000 bpd by the end of 1949. From 1950 through 1969, the amount of crude oil produced by ARAMCO rose at an average annual rate of about 9%. By 1970 daily average production had reached more than 3.5 million bpd. This rate tripled in the next decade and reached 9.5 million bpd in 1979. Thus, when Petromin first began operation in the 1960s, the petroleum industry in Saudi Arabia was huge by any contemporary standards. Exploration at that time was exclusively in the hands of ARAMCO, the Arabian Oil Company (AOC) and Getty Oil Company under the terms of the concession agreements.

The exploration activities are carried out in a fairly aggressive manner. Both the Ministry of Petroleum and Petromin are deeply involved in the evaluation of the exploration programme and they sometimes modify it for technical and logistical reasons. The programme is reviewed and updated periodically according to the new input from the exploration effort. Our role does not end at the evaluation of the exploration programme, however. It involves a continuous follow-up of the programme and its implementation and execution. The Ministry of Petroleum and Mineral Resources also carries out its own reconnaissance and exploration activities in the open areas to assess potentials and identify new prospects.

Now, a brief word on some of Petromin's activities. Between 1965 and 1968, Petromin participated in an agreement with AUXIRAP, a French company, and TENNECO to explore for oil along the Red Sea coast. In December 1967 the Kingdom conferred on Petromin two exclusive licences for hydrocarbon exploration in the Rub'al Khali and along the Red Sea coast and granted it the right to assign shares of interest in these concessions. Petromin entered into agreements with AGIP, Phillips and Natomas Sinclair (Pakistan).

These activities gave rise to a brisk exploration programme which produced useful exploration and survey data. Unfortunately, the effort did not result in any significant commercial discovery. Nonetheless, the experience was extremely valuable in the sense that we have learned at first hand the risks involved in exploration.

One other means by which Petromin sought discoveries remains to be discussed. The Geophysical Survey Company (ARGAS) and the Arabian Drilling Company (ADC) are two joint ventures between Petromin and reputable foreign companies in exploration and drilling work. ARGAS explores through geophysical techniques for oil, gas, minerals and underground water. ADC was established in 1964 to undertake all kinds of drilling operations both inside and outside the Kingdom. This kind of arrangement is another way of providing Saudi nationals with theoretical and practical training in modern methods of exploration and drilling.

From these facts it is clear how Petromin has assisted the Kingdom in reorienting the oil industry so as to serve national goals. These objectives, as I am sure you are aware, include massive economic and social development based on the Kingdom's hydrocarbon reserves. There are many ways in which developing countries might accomplish this restructuring of society. For Saudi Arabia the path has been a practical one and so the state petroleum enterprise in Saudi Arabia has pursued diverse paths in the exploration area to serve two major national interests — expansion of the resource base and acquisition of new technologies.

Petromin is keenly aware of its technical and manpower deficiencies. This has led it to explore several methods by which these problems could be resolved. Our first involvement has been as a supporter of educational institutions, the universities, such as the University of Petroleum and Minerals in Dhahran and the University of Riyadh, where geology, petroleum engineering and associated disciplines are stressed. Our second area of interest is in training centres. Petromin's and ARAMCO's training programmes for Saudi employees are only two examples of this effort. This educational effort now covers virtually all areas of exploration, petroleum engineering and other petroleum-related fields. However, both these training efforts require time and, until Saudi graduates are able to fill positions in exploration departments, the Kingdom will continue to need the services of highly qualified expatriates. We are acutely aware of the problems this temporary need presents. The worldwide pressure on technical manpower coincides not only with the staffing requirements of Petromin, but also with new demands created by the maturing of the Kingdom's oil fields. These, then, are additional reasons for the Government's and Petromin's interest in the establishment and staffing of the new exploration and petroleum engineering technical centre in Dhahran.

In conclusion, I should like to emphasize that, although the growing role of state petroleum enterprises in exploration, as in all other petroleum activities, is an irreversible evolutionary phenomenon, yet it does not mean that the involvement of the transnational oil companies, or of other companies in the field, will be adversely affected. With their experience and technical know-how, their positive and active involvement on acceptable business terms will always be welcomed.

The model of the role of a state petroleum enterprise in exploration that I have just presented to you has produced an adequately satisfactory result in meeting our planned objectives. I hope this presentation of our experience may be of service to other state petroleum enterprises in helping them to develop policies to suit their own particular circumstances.

15

STRENGTHENING NATIONAL TECHNOLOGICAL AND TRAINED MANPOWER CAPACITY IN PETROLEUM EXPLORATION: MEXICO'S CASE AND PHILOSOPHY

Eduardo J. Guzman

Adviser to the General Director, Mexican Petroleum Institute

The manner in which the exploratory effort in the search for oil and gas has been carried out in the different countries of the world has had almost as many variations as the number of countries in which this activity has been conducted.

Apart from the great differences resulting from the different times at which this exploration took place, there are many other factors which have accounted for the diversity in the methods used and policies followed and the magnitude and intensity of the effort carried out — and consequently of the results obtained.

Particularly worthy of mention have been, of course, the complex national, political, ethnic, economic, sociological and geographical characteristics of each country.

Although this paper is specifically concerned with petroleum exploration activities in the developing countries, it should be borne in mind, on the one hand, that there are many interrelated factors within the context of the worldwide petroleum industry and that, on the other hand, it is practically impossible to lump together in one category all of the so-called developing countries.

Since we all recognize that there is a wide range of conditions, it can be stated that each country — or region at least — has its own very distinct peculiarities. Accordingly, the advancement and development in the science or art of petroleum exploration in each of these countries has necessarily reached many different stages and is being approached in as many different ways as are dictated by the political, economic, sociological, educational and geographical settings of each country.

Mexico's experience in this field is intimately linked with the history of the Mexican petroleum industry, the way in which it has evolved and the manner in which it is now operated.

Since the latter part of the last century, several groups, some national but mostly British and American, undertook various exploratory projects in Mexico, attracted principally by the numerous natural oil seepages that were known to exist, mainly along the coastal plain of the Gulf of Mexico.

Among the men involved in this preliminary evaluation of the country's oil potential there were some Mexican mining geologists and engineers, but most of the reconnaissance work was done by Swiss, British, Dutch and, later, American geologists. At that time, petroleum exploration was a somewhat empirical endeavour, even for these experienced scientists, owing to the youthful stage of the oil industry and to the fact that little was known of what is now called 'petroleum geology'. The climatic and ecological factors required for the generation of oil, its behaviour in the subsurface as it migrates, accumulates and becomes trapped, together with the physical conditions and processes that produce commercial oil reservoirs, were at best mere theories, frequently mixed with myths and even superstitious beliefs. Furthermore, the tools available and the technological capacity of humanity to locate this valuable resource and extract it from the earth were still in their infancy.

Thus, in the early twentieth century, petroleum exploration, not only in Mexico but in the whole world, consisted basically in locating a natural oil seep and drilling or excavating a hole nearby. At that time, many now discarded theories, as well as some valid ones, were used to try and predict the presence of commercial oil in the subsurface. Among the latter was the anticlinal or structural theory which is still one of the most successful approaches to finding oil.

So, if we turn back only 80 years, we find our most experienced explorationists using all their knowledge in the simple task of locating and mapping anticlinal structures and in spotting surface occurrences of hydrocarbons.

It can fairly be stated that many sciences, such as geology, physics, chemistry, electronics, biology, palaeontology and many others, as well as the technologies that have stemmed from them, owe a great part of their rapid development to humanity's efforts to find and produce oil and gas. Unfortunately, we must admit that many of these advances have come also from man's natural aggressiveness.

The development of Mexico's technological capacity and trained manpower in this particular field of petroleum exploration has been a gradual process almost as old as the industry itself and cannot be attributed solely to Mexico's rebirth as an important oil-producing country. It is true that Mexico's new position has required us to redouble our efforts to strengthen our technological capacity and train specialized manpower but, fortunately, the seed had been planted years before, and so we can proudly say today that at least three generations of national petroleum explorationists have contributed to our present success.

In order better to understand Mexico's successful effort in this field, it is necessary to make a brief historical review of the Mexican oil industry.

Through the efforts of the first foreign and national pioneers, real commercial production was first started in Mexico in 1904, although some marginal production had been reported as early as 1901 and, on a smaller scale, even back in the 1880s. By 1921, Mexico had become the second largest world producer of oil, surpassed only by the United States. Production reached half a million barrels of oil per day, a considerable achievement at that time.

Mostly foreign capital, but even some national capital, had been attracted by the 1884 law which broke with the traditional legal theory that all mineral rights belonged to the nation. They were supposed to be a legacy from the Spanish Crown which became vested in the Mexican Government with the achievement of independence. The 1884 law somewhat resembled the laws of other countries, particularly the British Commonwealth and the United States, where ownership of land includes ownership of minerals beneath the surface.

The Constitution of 1917 rightly restored the ownership of all minerals and hydrocarbons to the nation but it permitted the state to grant concessions to individuals for the exploitation of these resources.

In 1938, owing to claims by the oil workers amounting to \$7.3 million per year and to a series of subsequent legal developments which are beyond the scope of this paper, the Mexican Government decided to expropriate 17 of the foreign petroleum and associated companies which were operating in the country. In 1962 Mexico paid the last of a long series of payments and wiped out its oil expropriation debt.

Forty-two years have elapsed since Mexico nationalized its oil industry and Petróleos Mexicanos (PEMEX) was born. The Government's decision to expropriate the oil companies was not premeditated, however much it may have been accused of premeditation by the supporters of private enterprise through the press of half the world. Yet, owing perhaps to the rebellious attitude of the oil companies in the years immediately preceding the act of expropriation, this was one of the most popular decisions the Government has ever made, which partly explains the enthusiasm surrounding the birth of Mexico's national petroleum enterprise.

Owing to Mexico's continuing insistence that its laws be respected and also owing to the international trend towards a greater socialization of industry, the companies undoubtedly felt insecure, and they gradually decreased their activity within the country, which led to a rapid decline in production. Many of the less well informed, both outside and inside the country, chose to believe that Mexico's oil potential was an exploded myth.

However, the real reason for the marked decrease in production during the late 1920s was that the larger, more powerful companies had almost completely ceased to explore and test new areas, and that Mexico's production had not been backed by sufficient proven reserves.

At the time of the expropriation, there were only a few Mexicans with enough technical background, but the challenge to the oil workers, the world wide scepticism about the Mexicans' ability to operate their oil industry successfully, together with the boycott imposed by foreign companies, acted as a powerful stimulus. Thanks to this, it was possible not only to keep the distribution of oil products going but also to ensure the manyfold growth of production, reserves, pipelines, refineries and — most important of all — to acquire experience. Of course, this could not have been done with patriotic enthusiasm alone. The effort would have failed had it not been for a handful of men who, undeterred by the long and difficult path that lay ahead, were determined to achieve technical self-sufficiency for Mexico's oil industry.

During the first few years after expropriation, the international events leading up to World War II had a definitely favourable influence on the development of Mexico's oil industry. Immediately after the war, a rapid expansion of the domestic market gradually converted an export industry into the most important source of energy in the country. Thus, the industry grew as a necessity of Mexico's development and at the same time contributed to the country's further industrialization.

These are the main prevailing background factors which explain the strong nationalistic spirit that reigns in the Mexican oil industry, which has become a tradition as sacred as the Act of Independence.

At the time of nationalization, there were just a few Mexicans who had acquired some experience, either with the foreign oil companies in Mexico or abroad, particularly in the field of exploration. They realized the immediate and future needs of the enterprise which was to be solely responsible for the development of the country's entire oil industry and for carrying out every operation properly, from the finding of the oil to the distribution of the products, without having to 'ask the Hague' (the phrase commonly used by the people who worked for the Royal Dutch Shell Group in Mexico, the Compañía Petroleo El Aguila, S.A.).

The needs to be met and the problems to be solved were many and they had to be dealt with in the short, medium and long term. The most pressing in the field of exploration can be summed up as follows: integration of all available information and knowledge; personnel training at all levels; acquisition and maintenance of the more advanced technologies; and, last but not least, the development of confidence in the scientific and technical competence of Mexican professionals, individually and collectively.

These four basic questions were, of course, closely bound up with other problems, such as economic capability, national and international policies and politics, management know-how, and so on, which it was sometimes more important to solve than problems of product distribution, labour relations, etc.

However, notwithstanding the delays and obstacles that the abovementioned problems may have caused, the basic steps towards a successful exploratory effort were readily undertaken, and it can now be said that they have produced fruitful results.

Full integration of the geographical, geological, geophysical and even historical data was of the utmost importance. Assembling and sorting all the information acquired and kept confidential by several competing and purely commercial enterprises was a gigantic endeavour, but when this had finally been done, the information became one of the most powerful tools with which to conduct, economically, an exploratory programme encompassing the whole of the country.

It may be possible to find oil faster under what is called private enterprise, which is so much vaunted in certain countries where competition, with everything that that implies, is almost a sacred duty, but it is done at the cost of squandering vast financial and human resources. Some of these countries have been able to afford it and still grow rich; but how can it be profitable to have the same tract of land surveyed three or four times by competing companies using expensive geophysical exploration methods, or to have wells drilled close to each other without knowing what was found in the previous ones — or, for that matter, to have four filling stations owned by different companies at the four corners of the same crossroads? One wonders how much might have been achieved if all these efforts had been pooled.

I am not a specialist in political science or economics but as a geologist, I should hate to have to work an area without knowing what there was on the other side of the fence.

Competition in all fields no doubt has its merits, but when dealing with a dwindling vital natural resource, Mexico's philosophy, embodied in the proposal our President, José López Portillo, submitted to the United Nations General Assembly in September 1979 — a proposal that all sources of energy should be the joint responsibility of all mankind — deserves more careful study by both oil-producing and oil-consuming countries. The Mexican proposal that a committee of the highest international authority — such as the United Nations — should be the regulatory body for the world's energy resources (and consequently for the petroleum industry) is truly humanitarian, in contrast to the purely commercial attitude of many countries, companies and individuals.

Criticisms have been heaped upon the national oil companies, particularly those of the developing countries, by the specialized international or transnational companies and in many cases by the governments of the countries the companies are supposed to represent or belong to. The national oil companies have been accused of inefficiency, bureaucracy, corruption, lack of organization, etc., which may be partly true, but these accusations are largely offset by the systematic integration and control of all the information and operations by any one country, which is, in the long run, a more powerful tool than the billions of dollars that the companies and their associated banks are able to bring into the field.

This has been very true in Mexico and we are certain that it can also be true in many other countries, but at the same time we must also accept the fact that many countries may not yet be ripe enough and should not try to run before they can walk.

Nationalism *per se* is not enough. Mexico had to learn and is still learning from others and must continue to do so, but thanks to our proud, clean international policies, we are able to pass on to others what we have learned, without expecting any substantial benefits in return, except a handshake and a friendly reciprocity.

The training of specialized manpower has necessarily been a long-term project.

During the first four decades of this century our universities produced only civil, mechanical, chemical, mining and petroleum engineers, so that most of the experienced personnel in the incipient exploration department started by PEMEX some forty years ago came from these basic disciplines. Geology was mostly in the hands of mining and petroleum engineers and geophysics in those of mechanical and electrical engineers. The first geologists that graduated in Mexico obtained their degrees in 1942. Petroleum engineers had been obtaining degrees since 1932 but most of them were busy on production and drilling operations. In exploration, as in other fields, there were, however, a few experienced technicians who helped the young and inexperienced professionals, and so gradually a team of geologists, geophysicists and biologists learning palaeontology, all with more enthusiasm than knowledge, began to lay the foundations for what was to be the future Mexican oil industry. Also, at the same time, a few Mexicans who had gone to college abroad or who had left with the companies after expropriation came back to reinforce this incipient exploration team.

It has been through scholarships and donations from PEMEX to the Mexican universities and polytechnical schools that a gradually increasing flow of geologists, petroleum engineers, electronic engineers and, more recently, geophysicists have begun to obtain their first degrees.

Some chosen individuals were sent to foreign schools to complete their education with master's and doctoral degrees, and although this has proved to be a good measure to some extent, since most of the high-ranking posts are now occupied by these men, many students never came back or else never readapted to their own country upon their return. This, as we know, is the case in many other countries as well, particularly those of the Middle East.

Although scholarships for study abroad are still being offered and should continue to be awarded, in more recent years it has proved to be more economical and more practical to bring the experienced professors to Mexico; thus, a growing programme of continuing education has been conducted, particularly since the creation of the Mexican Petroleum Institute.

The Mexican Petroleum Institute was organized as a research and development centre, and among its many projects in all fields of the industry one of those into which it has put the most effort has been the training of personnel at all levels.

The role of the Mexican Petroleum Institute in our national effort, which was conceived as strengthening and aiding the operational sector of the petroleum and chemical industries, has now become very important, although the Institute was originally started with a small group of experienced professionals trained in PEMEX, who still occupy most of the managerial posts.

Before the creation of the Petroleum Institute, the training of qualified personnel had to be undertaken solely by PEMEX and this was done through scholarships in geology and petroleum engineering at the national universities or polytechnical schools, thus producing an increasing number of graduates. In more recent years direct help to the academic institutions through donations has permitted the schools to offer master's and doctoral degrees in these disciplines, thus raising the level of the education that can be obtained within the country.

At the present time, some individuals are still being sent to foreign universities, but only on a very selective basis.

Since its establishment, most of the training has been provided through the Mexican Petroleum Institute, with an increasing number of continuing education programmes being offered. Since 1966 more than 5700 of these courses have been given.

The Institute also has a large division in charge of personnel training at all technical levels. Most of the training is given in the field by full time instructors. To date, more than 11 675 courses of this nature have been offered.

These educational and training programmes are directed to PEMEX's 117 000 employees, about 10 000 of whom are engineers of various disciplines, and to the 3500 employees of the Mexican Petroleum Institute, of whom about 1200 are professionals.

The acquisition and updating of technological capacity has mostly been secured through the employment of specialized consultants and the use of service companies, which, while operating in the country, have had the contractual obligation of training the Mexican personnel and using them in their work as soon as this was reasonably feasible. Today, although several foreign service companies still work in the country, particularly in geophysics, practically all the personnel is Mexican. Several Mexican nationals who started and were trained in the foreign service companies have established their own Mexican companies and now work for PEMEX or other institutions needing their specialized services.

In 1972, the Mexican Petroleum Institute established the first geophysical digital processing centre with the technical advice of a reputed American geophysical company and it is now operating around the clock to satisfy most of the processing needs of PEMEX. This has permitted better control and integration of the information and substantial savings in time and money.

Through somewhat similar arrangements, and also following the Government's guidelines, the Mexican Petroleum Institute has, in its short life, been instrumental in the acquisition and development of the latest technologies and has helped to promote the local manufacture of capital goods, particularly for the petroleum and chemical industries.

In conclusion, I wish to point to the strongly rooted admiration for imported goods and anything foreign that is to be found in many developing countries; the belief that experts must come from far away; the almost natural contempt for national capacity in technical matters — all of which are part of an attitude that prevails in particularly all of the developing countries. It was this attitude that prompted a group of Mexicans to bring an Austrian prince to govern Mexico in the 1860s. In Mexico we even have a word for this feeling, derived from the name of the Indian girl who joined Hernan Cortes and helped him to conquer her own people.

Fortunately, within the Mexican oil industry this lack of confidence in our own capacity has been offset by the nationalistic spirit that stemmed from the historical act of expropriation. We knew that we had to learn by personal experience alone and act on our own to the best of our ability. But this is not completely true in other developing countries or, for that matter, in other fields in Mexico, where foreign technologies or expertise are still considered essential even for relatively simple tasks.

Mexico has been fortunate in the development of its oil industry, although this has not been achieved without hazards; but many other countries still have a long path ahead of them. According to their own particular circumstances, they still have to acquire the self-confidence that is essential if they are to find and produce the oil to supply future generations.

PART VI

Bilateral and Multilateral Co-operation Programmes in Petroleum Exploration

16

PETROLEUM EXPLORATION IN DEVELOPING COUNTRIES: THE NEED FOR A GLOBAL STRATEGY BASED ON PUBLIC POLICY

Hasan S. Zakariya
Inter-regional Adviser, Petroleum Economics and Legislation, United Nations

THE CONTINUED BALANCE OF PETROLEUM SUPPLY AND DEMAND WORLDWIDE

The strategy for finding new petroleum reserves in the developing countries, though a legitimate issue in itself, would be more meaningful if considered and debated against the broader perspective of the future needs and potential of the world community as a whole. Such a strategy must not be contemplated in isolation, since it not only constitutes by necessity an integral part of any broader strategy encompassing the rest of the world, but also because the financial and technological assistance of the rest of the world will be crucial to the successful drive for petroleum exploration and development in the developing countries for many years to come.

Prior to 1973-1974, neither the continued availability nor the price of petroleum was a serious issue. Petroleum was both plentiful and cheap. If there was any real concern at that time, it was the new oil exporters' continued frustration in trying to obtain what they believed at the time to be the full value of their major source of national income. As from 1974, however, the exporters finally managed to obtain redress for what they had long viewed as their major grievance.

While for the last seven years or so the price of oil has been a source of varying degrees of serious concern to all the oil importers, the big importers, at least, seem to have reconciled themselves — in spite of public protestations and posturing to the contrary — not only to the sharp and sudden increases in the price of oil in 1974 and 1979 but also to what seems to be an irreversible trend upwards.

In fact it is the continued availability of petroleum at levels which can meet future world demand (even if anticipated growth over present levels were nil or modest) that is becoming the dominant issue, if not obsession. This is so in spite of the fact that, so far, oil has been available, at the asking price of course, not only for the current actual needs of the world community but also for hoarding and strategic stockpiling in the major centres of consumption.

This situation of almost perfect balance between demand and supply may last perhaps five, ten or even twenty more years (the exact time is immaterial considering the long-term perspective) but there will certainly be a time when an imbalance will begin to appear on the horizon. I do not have in mind temporary imbalances resulting from sudden disruptions of supply due to unforeseen circumstances affecting this exporting country or the other to be followed by panic buying — and even profiteering — as happened in 1979, for example. I am thinking rather of the time when there will be a genuine and more enduring shortage of supply resulting from the inability of available and produceable reserves to meet the needs of the world community.

When this will happen is, of course, uncertain; conflicting speculations and prophecies abound.[1] This writer is not qualified to play this highly uncertain but otherwise fascinating guessing game. Even if he were, he would be very inhibited indeed by the sobering sight of some past prophecies which are lying by the roadside like fallen leaves on a cold damp November day, trodden by indifferent or irreverent feet. The indisputable fact remains that petroleum is finite, non-renewable, non-recyclable and, at least for the foreseeable future, not easily replaceable by other sources of energy. Anticipated shortages could come about from one or a combination of factors:

(1) Growing concern among major exporters that their oil reserves are being depleted at an undesirably high rate. This has led some of them to restrict production in recent years in order to prolong the life of their reserves. This attitude is not confined to certain OPEC countries alone; other major producers, such as Mexico, Norway and the United Kingdom, for example, are also actively trying to maintain a sound ratio of reserves to production.

(2) The level of domestic consumption of oil within OPEC countries, until recently very modest, is steadily increasing owing to economic growth, increase of population and a rise in the standard of living. It is estimated, for example, that domestic consumption in Venezuela by the end of the century will reach 1 million bpd, and that if production is limited only to existing fields, the export potential of Venezuela will disappear entirely. Even if Venezuela were able to develop all available conventional petroleum deposits, a maximum production of 1.7 million bpd would be attained, leaving only 200 000 bpd for export, as against 1.8 million now.[2] In Algeria, to cite another example, it is estimated that the proportion for domestic consumption will rise from the present 21% (16 million TOE out of 75 million TOE) to 30% in 1990 and 77% by the year 2000.[3]

(3) It was anticipated that higher oil prices would reduce the rate of

consumption in the major industrialized countries, either spontaneously or because of the adoption of government regulatory measures. So far, this has been only partially and unevenly successful. The prospect for more stringent conservation measures in those countries does not seem very encouraging at the moment. In some developing countries also the price of oil products for domestic consumption is being kept unreasonably low, which is causing harmful waste.

(4) Higher prices were also expected to accelerate the process of searching for alternative sources of energy. However, the outlook for an early breakthrough in this regard is not as promising as the initial forecasts indicated. None of these new sources (wind, solar energy, etc.) is expected to become effective until perhaps after the turn of the century.

Measures to avert or temper anticipated shortages

As things stand now, it appears that, until the turn of the century at least, only new petroleum reserves can avert the probable shortages ahead. In order to preserve the present approximate equilibrium between world supply and demand, a global strategy would seem imperative for attaining some, and preferably all of the following objectives:

(1) Rational utilization of existing proven reserves worldwide. To some extent, this process has already started with varying degrees of success. More should be done, of course, in this respect. What is needed is some broad understanding between the major oil exporters and importers regarding the desirable rate of production and consumption and sincere co-operation in abiding by such an understanding.

(2) Re-evaluation and improvement of the rate of recovery of existing proven reserves. It may be said that, at present, only 25% (or at the utmost 30%) of such reserves can be extracted, but there has recently been considerable progress in the development of secondary and tertiary recovery methods. Their large-scale application, though costly, is expected to raise the recovery rate to 40% by the end of the century, and perhaps to 45–50% eventually.[4]

(3) Exploration and development of previously high-cost sources of petroleum such as the heavy oil deposits of the Orinoco Belt in Venezuela and the oil shales and oil sands in North America.

(4) Intensification of the drive to locate and develop new reserves worldwide both on land and on the continental shelf, including its outer limits. This process is, of course, already under way on a large scale in the industrialized countries of both the developed market economies and the centrally planned economies. To a lesser extent, it is also taking place in the more advanced of the developing countries, including the OPEC countries. Regrettably, the same is not true in the case of the oil-importing least developed countries.

The case of this group of countries is the principal concern of this paper.

HOW MUCH EXPLORATION IN THE DEVELOPING COUNTRIES?

As in several other respects, when it comes to discussing the scope and intensity of exploration activity, the developing countries are a very mixed bag indeed. The situation within the group is so varied that to attempt to apply one common denominator would be both unwise and unrealistic. On the one hand, there is that group of countries, mainly OPEC but including also some non-OPEC countries, whose oil potential has long been established and whose present reserves stand perhaps at more than 50% of total world reserves. On the whole, the full potential of this rather small number of countries is far from being fully exhausted and the prospects for further discovery there are still very promising. In those countries, exploration is being continued with varying degrees of intensity by state-owned companies, by foreign companies or by both combined. On the other hand, there is that other group of countries whose potential was thought to be so poor that little or no serious exploration has been carried out in them; what has been done to date has been only half-hearted and no attempt has been made to take a second and more scrutinizing look. In between, there are those countries where exploration activity has been carried on with varying degrees of adequacy and, in some instances, oil has actually been discovered and produced; nevertheless, their full potential is far from being conclusively established.

Whether oil exploration in the developing countries as a whole has so far been adequate or not is a matter of contention among different commentators:

(1) According to a report prepared by Exxon in 1978 (based on a poll of 29 experts at the World Energy Conference, 1978), exploration in the developing countries has been adequate on the whole. Fifty per cent of the exploratory wells (6501) drilled in the ten-year period 1967–1976 were in the non-OPEC developing countries. The share of the industrialized countries, excluding the United States, Canada and the centrally planned economies (CPEs), was 24% and that of OPEC 26%. Of the 113 non-OPEC developing countries where exploration (including drilling) was carried out in that period, 50 countries showed no encouraging prospects, 9 countries showed some encouraging prospects but no actual discovery and 20 countries witnessed no activities at all.[5]

(2) In contrast to the above, there are those who maintain that the level of oil exploration in recent years in the non-OPEC developing countries has been far below what is both desirable and feasible. South East Asia, Latin America, and Africa in particular, have been affected by a decline in exploration efforts. In the United States,

including offshore areas, they have shown a definite increase. In a recent study conducted for the US Geological Survey, Latin America and Africa are considered 'sleeping giants' in terms of petroleum potential. However, they are regarded as virtually 'undrilled' compared with the United States, for example. Of the more than 3 million exploratory and development wells drilled worldwide so far, 2.2 million have been drilled in the United States alone, which contains only about 10% of the world's prospective areas. In contrast, only about 100 000 wells have been drilled in Latin America, which is 1.6 times larger than the United States. The disparity is even more striking in Africa, where less than 50 000 wells have been drilled.[6]

The preponderant role of the private international sector

Historically, the initiative in the search for petroleum deposits in the developing countries has so far been left to the private international sector. There have been very few countries which have managed to accomplish the task alone with their own limited financial and human resources, and fewer still which have been able to persist until the end in maintaining an exclusive role for the state and its agencies. Mexico is perhaps the unique exception. Most of the others that opted at one time or another for complete and exclusive state control in this area have had to relax (of late) their policy against the involvement of foreign companies, allowing them to participate, with their enormous financial resources and sophisticated technology, in the mounting national drive to locate and develop indigenous petroleum resources. Witness, for example, recent developments in Brazil, India and even China, to cite only a few cases.

In the case of the least developed countries, the role of the private international sector has been even more decisive. In the absence of any alternative outside assistance, and lacking even the limited advantages enjoyed by some of the richer and more developed among the developing countries (in terms of natural resources, trained manpower, etc.), the least developed countries have been left completely dependent on the initiative and the interest of the foreign companies. When they demonstrate such an interest, the foreign companies are, of course, in a much stronger bargaining position than the government concerned. Because of their preponderant position worldwide and the lack of other sources of outside assistance, the deals these companies are prepared to offer are usually designed not only to secure the largest margin of profit and the recovery of total investment within the shortest possible time, but are also governed by such additional considerations as the cost of unsuccessful operations elsewhere and the need to hedge against political risks, real or imaginary, involved in dealing with the least developed countries. Since the risk involved in oil exploration is of special significance, over and above the normal risks involved in doing business abroad, we shall deal with it at some length.

THE RISK AND THE REWARD

Normal, inherent risk

Petroleum activity undoubtedly involves a certain degree of financial risk. The first and most obvious one is the inherent risk in the exploration stage. The operator, whether a foreign company or a national entity, stands to lose substantial amounts of money should the search prove futile. As far as the activities of the transnational companies are concerned, this has not been a major inhibiting factor over the years. Petroleum exploration has been a calculated gamble played by professional, cautious and consummate players who have at their disposal an array of accumulated experience and steadily improving technology to minimize the incidence of failure and increase the likelihood of success. Besides, whatever the inherent risk in any particular venture, the resulting loss has always been compensated for by success in other ventures either within the same country or elsewhere. For the major oil companies, the stage is the world. On balance, the incidence of success has been much greater worldwide than that of failure, and the professional gamblers have been growing more and more prosperous.

Political risk

There is of course another type of risk in petroleum ventures in developing countries quite unrelated to the nature of petroleum exploration itself, but resulting rather from the political and socio-economic conditions in any particular country at any given time. This tends to engender fear of total or partial loss of investment after the discovery and production of petroleum, or at least the erosion of initial financial expectations.

It is, of course, understandable why companies are reluctant to venture into a country where there is evidence of political instability or where the government subscribes to a doctrine hostile to foreign or private enterprises. But this is not peculiar to petroleum; it is part of the normal hazard of doing business in foreign lands.

In practice, however, political risk has not had a great or uniform impact on the behaviour of foreign companies. Normally they all take into account the so-called 'investment climate' and its suitability to their objectives before deciding on an investment venture. In the words of none other than Exxon itself,

> The assessment of an unwillingness to accept political risk varies among companies, apparently in direct relation to their estimates of geological attractiveness . . . A survey of current industry exploration activities indicates that there are few countries where political risk is a totally insurmountable factor. . .[7]

The attitude of the oil companies in this regard would seem to be no different from the general conduct of all the multinationals. As was observed in one newspaper article:

> Multinational corporations do not intend to give up the money game just because the world has become so politically hazardous. The returns in many unstable areas still look rich if money can be made there fast enough.[8]

One particular type of political risk about which the oil companies are apprehensive relates to a unilateral change in the contractual relationship short of complete nationalization or expropriation — what is called 'changing the rules of the game'. In justifying their reluctance to operate in some developing countries or their insistence on an unreasonably high reward for such risk-taking, the companies point out the unreliability or irresponsibility of the developing countries in general, as demonstrated by their failure to abide by the terms of their agreements — their attempt to change the rules of the game after the play has begun.[9]

The truth is that the occasional demand of developing countries for the renegotiation of previously agreed petroleum or mining contracts does not flow from their greed or delinquency, but rather from the circumstances under which nearly all such contracts were negotiated. The old concession agreements concluded during the colonial era are of course no longer at issue. These have now become relics of the past, since they have nearly all been liquidated or restructured. What is at issue is the current or recent contracts into which the host countries entered quite voluntarily, and apparently with sufficient awareness of what was at stake. No matter how well informed and how astute and steadfast a developing country — or any other country for that matter — may be in negotiating a petroleum exploration contract, it is impossible to predict with any certainty the shape of things to come over 30 or 40 years, which is now the normal life span of such contracts. Depending on whether petroleum is found and if so in what quantities, the exact cost of developing and extracting it and its price, what constitutes a fair and reasonable reward to be allotted to the contractor could vary considerably from one set of unforeseen circumstances to another. There are so many other imponderables. From the first petroleum discovery to its full reappraisal and development, there may be a period of 10 years and sometimes even longer. From the time the field is rendered productive to the expiry of the contract — or the complete exhaustion of productivity, should that occur first — there is usually, under current practice, a span of 20 to 30 years. Any attempt to predict what the situation will be over such a long period of time must be open to serious error on both sides of the contracting table. At best it is the exercise of an approximate wisdom, at worst, a game of guess work, if not a pure gamble.

It must be pointed out that a unilateral change in the rules of the game has not been the exclusive sin of the developing countries. Some of the oil producers among the developed industrial countries seem of late to have been guilty of this also; witness the case of the North Sea, where the Governments of both the United Kingdom and Norway have been repeatedly changing the contractual fiscal regime by introducing new taxes and restricting production to levels deemed to be in the national interest. More recently, the Government of Canada has embarked on a new energy programme which would entail a major restructuring of the petroleum industry in a way certainly not foreseen or authorized by the contracts held by the foreign oil companies operating there. One important aspect of the new policy, characterized by the foreign companies as 'confiscatory', is the Government's demand for a 25% 'carried' interest in all non-producing exploration ventures on federal territory without compensating the lease-holders for past risk expenditure.[10]

Contracts freely entered into should, of course, be faithfully observed and implemented so long as it is reasonably equitable to do so. Otherwise, contractual terms will have to be revised and readjusted, with mutual consent if possible, to reflect changed circumstances. In order to minimize future conflicts arising from changing the rules of the game it would seem desirable to make the rules less rigid and more open-ended, by agreeing on changes in advance and on a suitable mechanism for periodical review and readjustment to be stipulated in the contract itself, the paramount goal being to ensure that the economic rent under all petroleum development contracts should belong to the state, the lawful owner of these resources.

The high premium of risk taking

While the foreign companies, as a rule, are not unduly inhibited by the prospect of the various risks involved in operating in developing countries, the reward they demand in return for assuming the inherent risk of oil exploration and insuring against political risk is usually high, exceeding the bounds of a fair return on investment. To that end, not only do they demand the most favourable fiscal arrangements and the least onerous obligations, but they also try to include in the contractual package — usually drafted in minute detail — clauses which would make it almost impossible for the host government to increase its share from future additional benefits not envisaged initially in the contract. Clearly this is not the way the companies deal with any of the industrialized countries of the West. As has rightly been pointed out:

> A mining or oil company would get short shrift if it sought from the British Government contractual guarantees binding Parliament for the future in terms of a specific fiscal package. Developing countries may have to swallow their pride and accept contractual limitations on their legislative competence. However, if some instability results, it should not be seen as something rooted in 'politics'. It is better to be seen as a consequence, perhaps inevitable, of the kind of agreements which developing countries are obliged to accept.[11]

This situation would confront the developing country with an extremely hard choice: either to refuse the kind of deals offered by the companies with no alternative assistance in sight, or to accept grudgingly what is offered in the vague and pious hope that the unbalanced deal will somehow be rectified in the future.

To relieve these countries of such a dilemma, an attempt is made in the next section to inquire whether there is, or ought to be, another source of outside assistance that could replace the traditional role of the private international sector in oil exploration or, at least, underwrite the normal risks involved in such exploration and thus eliminate or reduce the highly inflated premium placed by the foreign companies on such a risk. Inevitably, this will lead us to review the present role of the public international sector and consider whether this role ought to be expanded and intensified.

THE PRESENT ROLE OF THE PUBLIC INTERNATIONAL SECTOR

The difference between the public and the private international sectors lies, of course, not only in the fact that the funds provided by the former are publicly owned, but also that, in extending such financial assistance, the public sector is, or ought to be, motivated by notions of public policy and not by profit-making. This is generally true even though some donor states may be moved also by certain political considerations of their own which would render their continued aid to some extent conditional upon the realization of certain expectations. These anticipated benefits do not, of course, amount to the kind of profit-making that underlies the activities of the private sector, and they are not usually stipulated or spelled out in the aid agreements.

As far as financing petroleum exploration and development in the developing countries is concerned, the role of the public international sector has been, regrettably, very limited, although its potential in this regard is considerable, if not decisive.

International institutions

The United Nations and affiliated bodies

For nearly two decades now, the United Nations Development Programme (UNDP) has been an important source of funds in the search for and development of natural resources. However, the bulk of this financing has been in the sector of hard minerals (non-fuel): exploration for petroleum has rarely been included. During the period 1960–1976, for example, the total expenditure on mineral exploration projects sponsored by UNDP was $168 million, of which $103 million was provided by UNDP and the remainder by host governments.[12]

The UNDP Revolving Fund

In 1973, the United Nations established the Revolving Fund for the Exploration of Natural Resources, to be administered by UNDP. Although the Fund was originally conceived to apply to all natural resources, the Governing Council of UNDP decided that the Fund should concentrate its initial operations on solid minerals only, thus excluding, among others, all petroleum resources. Early in 1980, an attempt was made by UNDP to establish an energy fund for seismic survey and other similar petroleum-related activities with administrative arrangements similar to those of the Revolving Fund, but no definite action was taken at that time. Instead, a group of government experts convened by UNDP decided to support in principle a horizontal expansion of the Fund into oil and gas exploration, provided it was limited to pre-drilling survey works, and that substantial additional funding became available. The function of the Fund in the area of petroleum was conceived to be mainly a means of reducing the risk factor by providing developing countries with additional geological data and thus enhancing their ability to attract more petroleum financing from the international private sector.

The World Bank

Until very recently it was the policy of the Bank not to assist in or finance projects for petroleum exploration and development. That policy was apparently due to the conviction that such projects are highly risky and that they normally require considerable funds. It was no doubt also encouraged by the fact that petroleum was not only dirt cheap but was also in ample supply worldwide. If we add to that the availability of well-established sources of risk capital represented by the major and other private companies which were prepared to undertake that task all over the globe, we can understand why the World Bank was reluctant for so long to venture into the field of petroleum.

All that seems to have changed with the so-called 'energy crisis' of the mid-seventies. In July 1977, the Executive Directors of the Bank approved a programme calling for an expansion of lending by the Bank Group in the development of the fuel and non-fuel mineral resources of member countries.

The new policy was designed to be limited in scope and supplementary in nature with greater emphasis on the production aspect of petroleum operations, which is normally free of risk. However, it was accepted that

> in a *limited* number of cases, the Bank might provide *small loans* for exploration to be refinanced by a subsequent production loan, or repaid by the foreign oil company if a commercial field was not found. The Bank would express its willingness to consider financing the production facilities if the exploratory effort was successful.[13]

It is interesting to note how this new lending facility was actually put into operation. According to its most recent report, the Bank has so far financed 18 petroleum projects in 16 developing countries. Its lending budget for petroleum alone during the fiscal period 1979-1981 will amount to $1.35 billion. Forty-nine developing member countries have been visited by Bank missions, and in 38 of these one or more projects have been identified. Of the 18 projects already financed, 9 are mainly for development activities, including technical assistance, geological and geophysical surveys and exploratory or appraisal drilling. The other nine projects are mainly for production.[14]

Petroleum Exploration: Compared to pre-exploratory drilling and the development and production of a discovery, petroleum exploration *per se* (with all its hazards and risks) has not fared well so far. In those few cases where Bank funds were made available, the main attention has been focused on countries like Morocco, in order to 'complete the exploration of a basin in which discoveries have been made but have not yet attracted foreign oil companies', or Tanzania, where an 'existing lead has been relinquished by an international oil company, either because the prospect was not judged to be economical at the time or for other reasons'.[15] It is both interesting and revealing that among the other 20 developing countries in Asia, Africa and Latin America which the Bank named or identified as possible candidates for future similar assistance at least 15 are petroleum-producing countries with experienced national oil companies, and two-thirds meet the Bank's own criterion of 'middle-income countries'.[16]

Bilateral assistance for economic development

OECD countries

The governments of the Organisation for Economic Co-operation and Development (OECD), encompassing all the developed industrial countries of Western Europe, North America and Japan, have been for some time now providing assistance to some developing countries on a bilateral basis in the form of loans and grants for economic development.[17] Except for a few recent instances such as, for example, the assistance given by Norway to Jamaica and Canada to Nicaragua for petroleum preliminary surveys, the funds they have provided have seldom been directed towards petroleum exploration in the recipient countries. Apparently this was due to the same reasons that led the World Bank, for example, to refrain, until three years ago, from granting loans for such activities.

There are, however, indications that the thinking of the OECD countries on this matter is beginning to change. They seem to feel now that the task of finding new petroleum reserves in the developing countries cannot, and perhaps should not, be left entirely to the initiative and the interest — or lack of it — of the private international sector.

No discussion of the present or potential role of the OECD countries would be complete without, however, discussing their oil companies, which for decades now have had a leading role on the world stage.

The so-called Seven Sisters belong to two countries — three, if one counts the Dutch share of Shell. Six of these are completely privately owned. One of them — BP — is half-owned by the British government. The French CFP, which could be counted as an eighth sister, is also partly owned by the French government. There are dozens of other privately owned companies — American, Canadian, Japanese and others — which have also been active for many years now on the international scene.

In addition to the above, there are several national oil companies which are either wholly owned by the state, as the Italian ENI and the recently established Japan Petroleum Development Corporation, or majority owned, as the French ELF/AQUITAINE. There are also private oil companies which are actively sponsored and supported by their home governments, like the Deminex of the Federal Republic of Germany. Historically, these state owned companies were created as an instrument of national policy of their respective governments, not only to search for and develop local petroleum deposits, if any, but, more importantly, to venture abroad and try to obtain some degree of security of foreign supply, earning on the side other economic and commercial benefits.

These state owned or sponsored companies of the OECD countries have, naturally, been mandated in their charters to operate in accordance with the usual norms of business — profit-making being their primary objective. When they first appeared on the international scene (in the 1950s and 1960s), they tried, for tactical reasons of their own, to offer the developing countries with which they wanted to negotiate new petroleum deals. These were somewhat more favourable to the governments concerned than the old concessions under which the major companies were then operating, both from the standpoint of the doctrine of state sovereignty over natural resources and in terms of the national share of benefits under the fiscal regime. Thus they succeeded in gaining a foothold in some of the promising areas and the goodwill of their host governments.[18]

Now that the old concessions in the developing countries have been completely liquidated or restructured and the major companies are prepared to play (and are actually playing) the game according to the same new rules introduced during the last quarter of a century by the state owned and sponsored companies of the OECD countries, there no longer seems to be any perceptible difference between the way the two groups of companies have been conducting their business of late in the developing world.

Countries with centrally planned economies

Some of the governments of this group of countries — particularly the Soviet Union, which possesses long experience in the field of petroleum exploration and development — have been involved, through their own state enterprises, in assisting some developing countries to locate and develop their indigenous petroleum resources. At one time or another, the Soviet Union has provided financial and technical assistance in petroleum exploration to Afghanistan, India, Iraq and other countries in the Middle East and Africa. Normally, such aid takes the form of loans to be repaid, with or without entitlement to the right to purchase certain amounts of the oil or gas produced. Full data on the modalities of such bilateral co-operation in the petroleum sector and the terms and conditions thereof are regrettably not readily available to this writer at the moment.

Co-operation between developing countries

It is worth noting that in recent years there have been an increasing number of instances involving the co-operation, on a bilateral basis, between the governments of developing countries themselves in the field of petroleum exploration and development.

Thus the Oil and Natural Gas Commission of India (ONGC) has carried out exploration work at one time or another in Iran, Iraq, Syria, and Tanzania, the National Oil Company of Argentina (YPF) in Uruguay, Petrobras of Brazil in Colombia (where it is already producing), Algeria, Iraq (where it has made significant discoveries), Egypt and Libya.

More recently, Tanzania was reported to have signed an agreement with Algeria whereby the Algerian state oil company SONATRACH will explore in the offshore area of the Song Song Island, 170 miles south of Dar es Salaam, where drilling financed by the World Bank has already located gas reserves and some indications of oil.

Elsewhere, Mexico and Venezuela jointly agreed to guarantee nine Central American and Caribbean countries with 160 000 b/d of crude oil on concessionary terms. These countries will have the option of receiving 30% of the cost of the oil returned in the form of a five-year loan at 4% interest, to be reduced to 2% if the proceeds are used for development projects, with priority given to developing indigenous energy resources. In this connection, it was recently announced that Pemex of Mexico and Costa Rica's state owned refining company have signed an agreement under which Pemex will explore for oil in Costa Rica in the Limon Basin.

Still in Latin America, the oil Ministers of the Latin American Energy Organization (OLADE) approved recently a Venezuelan proposal for regional co-operation in energy development. The Ministers agreed to establish a special fund to finance energy development in the region's poorer countries.

THE ATTITUDE OF OPEC

Before discussing the attitude OPEC countries should take regarding any international strategy for petroleum exploration in other developing countries, it is useful to inquire first whether or not such a strategy would also be in the best interests of the OPEC countries themselves.

There was a time not long ago when the oil producers and exporters grouped in OPEC used to view with apprehension, if not real alarm, the efforts made to discover new reserves in other parts of the world. At that time their fear was, of course, understandable, if not somewhat exaggerated. On the whole, their existing proven reserves were ample, the price of oil was low, and consequently the only way for these countries to increase their oil revenues was to expand production, even at the expense, on occasion, of their fellow OPEC members. They viewed the discovery and development of new reserves elsewhere as a real threat, a double jeopardy not only because such reserves tended to idle part of their production capacity, but also because of the negative impact the reserves might have on the price of oil. The OPEC countries stood to lose precious revenues on both counts. This was particularly so because at that time those countries were not in a position, individually or collectively, to take protective measures to counter such a threat or minimize its effects. Both oil production and oil prices within OPEC were under the effective control of the major oil companies, which happened to be, more often than not, the very same operators that were responsible for finding such new reserves, as, for example, in the North Sea and Alaska.

Since 1973, the situation has changed profoundly and dramatically. The OPEC countries are now in full co-ordinated control of the price of oil. Although they have failed to agree on a collective strategy for administering production within the entire community, the control of production in each country has passed from the foreign companies to the government. Each government can now increase, reduce or readjust the level of production at will.

With this radical change in the role of governments, increasing oil revenues has become a function of higher prices and not of larger production. Consequently, certain OPEC countries, like Kuwait, Iran, Libya and Venezuela, have reduced or frozen allowable production levels and have been tenaciously holding the line in spite of the temptation or pressure to produce more. Even with substantially reduced production, some major producers maintain that their current rate of production is still not only harmful to their own long-term national interests but also far in excess of their present financial needs. Oil in the ground, they believe, is better than the opportunities available to them now for investing surplus oil revenues.

It is submitted that the present policy of the OPEC countries to prevent undue depletion of their existing reserves is both wise and commendable, not only from the viewpoint of their own national interest but also in the long-term interest of the world community at large. As previously indicated, the same policy is being adhered to generally by major non-OPEC producers such as the United Kingdom, Norway and Mexico. No matter how large the petroleum deposits which are yet to be located and developed, either within the current producing areas or elsewhere, may be, this highly versatile, useful, but non-renewable material should not be allowed to vanish within the lifetime of one, two or even several generations to come. This being so, it would certainly seem to be in the best interest of OPEC itself, not only to view with equanimity the drive to locate new reserves worldwide but also to encourage and contribute to its success.

As far as petroleum exploration within the OPEC countries themselves is concerned, and in spite of occasional complaints by some to the contrary, efforts to that end are under way with or without the co-operation of foreign companies at what seems to be, on the whole, an adequate pace. These companies are, by and large, still demonstrating a keen, continuing interest in undertaking further exploration within OPEC countries even when the anticipated contractual benefits are rather modest compared to the 'good old days'. Moreover, some producers, like Algeria and Libya, for example, have been remarkably successful in devising some rather ingenious means to coax unwilling and reluctant foreign companies to undertake new exploration work.[19]

As to exploration efforts in other developing countries, particularly the least developed countries, we feel that OPEC should actively assist them in any possible way. Apart from the enlightened self-interest of the OPEC countries themselves, the discovery of new reserves in other developing countries would have yet another intrinsic value.

The bonds between OPEC and other developing countries are, of course, many and strong. OPEC countries have been demonstrating their concern for the welfare of other developing countries by eloquent deeds, individually, regionally or collectively, through the OPEC Fund for International Development (OFID). Although there might be room for further improvements in scale and quality, the economic assistance extended to the oil-importing developing countries has been impressive.[20] OPEC has, over the last few years, been willing to provide financial assistance, by way of outright grants or soft loans, to a great number of these countries either for non-oil economic projects or for normal budgetary accounts.[21] The question remains: Why not provide similar assistance for petroleum exploration which, if successful, would render the recipient country completely or partially self-sufficient? Could OPEC extend a better service to these countries, and perhaps indirectly to itself also?

It is very gratifying indeed to note the recent emergence of this line of thought within the ranks of the OPEC policy-makers, even though it has not yet gone beyond the confines of general sentiments and declared intentions. The first and the most important public expression in this regard was the draft 'Long-term Policy' formulated by OPEC's Long-term Strategy Committee, which is currently under consideration. The draft declaration presented to and discussed at the OPEC Ministerial Meeting held from 15 to 17 September 1980 in Vienna and which was to be finally submitted to an OPEC Summit Conference in November 1980, now postponed, includes the following:

> Furthermore, the Sovereigns and Heads of State, after examining the problem of the energy-deficient countries of the Third World,

> stress that their energy problems have to be seen within the general perspective of development, and they conclude that the *only viable long-term solution to the energy needs of developing countries is to develop the potential energy sources in their own countries.* Not only would this help solve the long-term impasse of developing countries, it would also contribute to improving world energy balances. In this context, greater emphasis should be placed on collective self-reliance among developing countries ......[22]

The draft declaration went on to propose the establishment of a long-term development fund to help energy-deficient countries solve their energy problems. The proposed fund, to be formed by contributions from both the industrialized world and OPEC countries, would be devoted to financing, among other things, 'development of indigenous energy resources in developing countries in a manner that preserves their independence, security and national sovereignty'.[23]

MULTINATIONAL FUND FOR UNDERWRITING NORMAL RISKS OF PETROLEUM EXPLORATION

It is indisputable that petroleum exploration requires large capital investments and a high level of technology and skills. It also involves considerable financial risks. It is likewise indisputable that most if not all of the oil-importing developing countries are in no position to meet all these requirements on their own. Even if they were, it is questionable whether they should divert their limited financial resources to ventures with such inherent risks.

Since the turn of the present century, the private international sector, represented mainly by the oil companies of the industrially developed world, has performed the task worldwide and has assumed all the risks involved, at a price. Technically and managerially, it has been, on the whole, a job well done. But the cost to the developing countries concerned has been for many years unreasonably high in terms of effective state control and the equitable sharing of benefits. In recent years some of these countries have managed, however, to rectify this anomaly by asserting state participation in and control of their petroleum industry, thus receiving most if not all of the economic rent derived therefrom. Once this was accomplished, it opened new avenues for fruitful and more amicable co-operation between these countries and the foreign companies.

While the case of the exporters grouped in OPEC against the private international sector can be considered closed, regrettably, the same is not true of other developing countries, particularly the least developed. In addition to a league of other economic ills, some of the countries in this group suffer also from what seem to be poor petroleum prospects which discourage the private international sector from embarking on or continuing with petroleum exploration. When they do accept the task, the private sector companies usually place a high premium on the risks they undertake, reflected in the rewards they insist on assigning to themselves in the event of successful exploration.

While the public international sector has recently recognized the urgent need of the developing countries to search for and develop whatever petroleum deposits lie under their territory, and has offered to assist them in this field, the actual assistance offered so far has been too little and too late. Even though they have a real long-term stake in increased petroleum exploration everywhere, the direct official assistance emanating from the OECD countries has been distinguished by its absence. The World Bank, which has been the most notable source of funds for economic development, is still unwilling or unable, with a few recent exceptions, to finance the most crucial of all petroleum operations: drilling exploration. When such financing is forthcoming, the limited funds provided are naturally enough considered loans to be repaid with interest by the recipient governments, even if the outcome is complete failure.

Among the various successive petroleum operations, drilling exploration involves the greatest risk. In practical terms, the big 'question mark' can be answered only by conducting a serious drilling programme. The preliminary geological and geophysical surveys are necessary and useful of course, but they do not usually cost very much and, therefore, the financial risk involved is minute. Once oil or gas is discovered and its commercial possibilities are ascertained, its appraisal and development require vast investments but likewise involve no major risk. The big gamble lies in drilling for exploration. The difference between a dry and a wet hole can mean many millions of dollars!

The financial risk of drilling exploration is the focal issue facing the least developed countries. Who is to sponsor this particular phase and underwrite its risk? The enterprises of the private international sector, which have traditionally undertaken the bulk of these operations, may be unable or unwilling in the future to continue shouldering the burden unless they are given additional incentives by way of risk sharing or larger profit margins.

In the light of the above, there seems to be a genuine and urgent need, as part of a global strategy for discovering more recoverable petroleum, to establish a new multinational fund to undertake, *as an international public policy*, the capital investment required for petroleum exploration in the developing countries, and to underwrite the financial loss incurred in case of failure. Without eliminating or replacing altogether the conventional role of the private international sector in this regard, the new multinational facility would act as a parallel or complementary channel. The two sectors could co-operate jointly together with the developing country concerned, the private sector companies providing for a fee all the technical services required without being responsible themselves for undertaking the risk involved.

Should the drilling exploration fail to discover commercial oil or gas, the loss of investments would be borne by the proposed multinational fund alone on behalf of the developing country. Should there be such a discovery, the country would be required, after production begins, to pay back into the fund all the investments incurred, with suitable interest. In addition, it might also be asked, in certain circumstances, to contribute extra money by way of a bonus to help replenish the fund and so enable it to extend similar useful services to other developing countries.

It is assumed that the initial capital of a multinational fund for petroleum exploration in the developing countries and its recurrent replenishment would be subscribed to by all the nations in a position to do so, primarily the OECD and OPEC countries. The latter group has already signalled its general readiness to do so. It remains to be seen whether the former will also rise to its international responsibility and meet the challenge of this worthy cause.

It is, of course, premature to discuss and articulate, at this juncture, the various formation and administration aspects of the proposed fund. Once its establishment is accepted in principle, the details can be worked out by those concerned on an *ad hoc* basis within the so-called North/South dialogue, and within the general framework of the United Nations or any other appropriate framework. Matters to be debated and settled would naturally include the following: ratio of contribution to the initial capital and subsequent replenishment thereof from the two above-mentioned groups of countries; whether to set up completely new machinery or integrate the proposed fund with existing international lending facilities like the World Bank and its affiliates; which developing country should have priority in calling for assistance from the fund; what the exact role and rewards of the private international sector should be in vigorously carrying out the basic aims of the fund; whether or not recipient countries should be made to contribute to the fund, in the event of successful exploration, anything over and above the loans provided.

NOTES AND REFERENCES

1. The peak for world oil production, excluding that of the CPEs, was projected, for example, as early as 1985 by British Petroleum and 1990 by Shell. On the other hand, Peter Odell and Kenneth Rosing, in their book *The Future of Oil*, contend that a world shortage could be averted for at least 30 years and possibly longer. See *Petroleum Intelligence Weekly* (New York), 3 November 1980, p. 2. See also the article by Pocock, C.C., 'Prospects for oil and gas: a look ahead to the year 2000', *World Oil* (Houston), October 1979, pp. 107-111, and 'BP draws gloomy reserves picture', *Oil and Gas Journal* (Tulsa, OK), 14 July 1980, p. 58.
2. *Platt's Oilgram News* (New York), 3 December 1980, p. 1, citing a statement made by the Director of the Venezuela State company, Lagoven.

3. *Arab Oil and Gas* (Paris), 1 December 1980, quoting an official Algerian source. See also 'Domestic energy requirements in OPEC member countries', a country-by-country analysis of the projected energy needs of the 13 member countries up to the year 2000, published by the Public Information Department of OPEC, August 1980.
4. 'How much oil in the World?', *Petroleum Economist* (London), March 1978, p. 86.
5. Exxon Corporation, 'Exploration in developing countries', June 1978, p. D4 A27.
6. *Oil and Gas Journal*, 8 March 1976, p. 32, quoting Bernardo Grossling, who conducted the research study. Grossling cautions that the number of wells drilled is not a direct indication of the amount of oil that will be found, giving as an example the case of the Middle East, where fewer than 50 000 wells have been drilled but large amounts of oil have been found.
7. Exxon, *op. cit.*, p. D4 A12.
8. Article by Leonard Silk, *The New York Times*, 14 February 1979, p. D2.
9. Faber, M. and Brown, R., 'Changing the rules of the game: political risk instability and fair play in mineral concession contracts', *Third World Quarterly* (New York), January 1980, p. 104.
10. *Petroleum Intelligence Weekly* (New York), 24 November 1980, p. 4.
11. Faber and Brown, *op. cit.*, p. 113.
12. United Nations. Multilateral development assistance for the exploration of natural resources. Report of the Secretary-General (A/33/256), 16 October 1978.
13. World Bank. A programme to accelerate petroleum production in the developing countries (E/1979/93), January 1979, p. 6.
14. World Bank. *Energy in the Developing Countries* (August 1980), p. 68.
15. *Ibid.*, p. 69. Into this category falls also the loan of $25 million which the Bank is extending to Egypt to enable its General Petroleum Corporation (EGPC) to carry out seismic surveying and exploration drilling in the Western Desert. See *Platt's Oilgram News*, 12 December 1980.
16. *Ibid.*, p. 69. See also pp. 74-75 and p. 79 (annex I).
17. The aid, which was never spectacular to start with, has been decreasing in recent years. Robert McNamara, the President of the World Bank, has termed the Official Development Assistance (ODA) extended by the industrialized nations to the oil-importing countries a miniscule and insignificant fraction of the gross national product. See his address to the joint meeting of the World Bank and the International Monetary Fund in Washington on 30 September 1980. *Middle East Economic Survey* (Nicosia), 20 October 1980 (Supplement), p. 6.
18. For further details see Zakariya, H., 'New directions in the search for and development of petroleum resources in the developing countries', *Vanderbilt Journal of Transnational Law*, Vol. 9, no. 3 (summer 1976), pp. 545–577, and 'State petroleum companies', *Journal of World Trade Law*, Vol. 12, no. 6 (November/December 1978), pp. 481-500.
19. Until September 1980 Algeria, for example, aimed at making participation in exploration a prerequisite for crude oil supply contracts. It was charging oil customers a temporary reimbursable premium of $3 a barrel in lieu of exploration expenditure, which was lifted when the customer concluded a new deal for exploration in Algeria. *Petroleum Intelligence Weekly*, 15 December 1980, p. 6.
20. In his address of 30 September cited in note 17 *supra.*, Robert McNamara characterized the Official Development Assistance of the six capital-surplus

OPEC countries — Saudi Arabia, Kuwait, Iraq, the United Arab Emirates, Libya and Qatar — as remarkable, pointing out that they contributed 4% of their combined gross national product during 1974-1979. In contrast, the foreign aid of the United States in 1979 is reported to have amounted to only 0.19% of the gross national product (dropping from 0.5% in the Eisenhower/Kennedy years). See the *New York Times* editorial 'Learning how to count foreign aid', 4 February 1981, p. A2.

21. OFID financial commitments have reached some $1.5 billion, including $910 million in loans to 76 developing countries, and the rest in grants and contributions to other development agencies. *Middle East Economic Survey*, 19 January 1981.
22. *Arab Oil and Gas*, 16 October 1980, p. 28. A few months ago, *The Economist* of London asked rhetorically 'Will OPEC finance an institution designed to help break the oil stranglehold?'. OPEC, apparently, does not mind! See 'If OPEC pays the piper', *The Economist*, 6 September, 1980.
23. See *Arab Oil and Gas*, *op. cit.*, note 22 *supra.*

LIST OF PARTICIPANTS

Member states of the United Nations

Angola

Arnaldo Lago, Financial Director, SONANGAL, The State Angolan Oil Company, Luanda.

Antonio Mangueira, Assistant Director, The State Angolan Oil Company, Luanda.

Argentina

Francisco Saraniti, Geophysics Manager, Yacimientos Petrolíferos Fiscales, Rogue Laenz Peña 777, Buenos Aires.

Barbados

Leslie H. Barker, Chief Geologist, Energy and Natural Resources Department, Ministry of Trade and Industry.

Botswana

Colin Clark, Director, Geological Survey Department, Private Bag 14, Lobatse.

Brazil

Raul Mosmann, General Manager for Exploration, Petróleo Brasileiro S.A.—PETROBAS, Av. República de Chile, 65/713, Rio de Janeiro—RJ—20035.

Bulgaria

Georgi Manchev, Second Secretary, Bulgarian Embassy, The Hague, The Netherlands.

Burundi

Liboire Muhagaze, Chef du Service de Géochimie, Département de la Géologie, Ministère de l'Energie et des Mines, B.P. 745, Bujumbura.

Canada

Ian MacDonald, Export Director, Consulting Services and Construction, Department of Industry, Trade and Commerce, Ottawa.

Robert Meneley, Group Vice-President, Exploration, PETRO-CANADA, 350 Sparks Street, Ottawa, Ontario K1R 7S8.

S.A. Rayer, Chief Geologist, PETRO-CANADA, 350 Sparks Street, Ottawa, Ontario K1R 7S8.

Robert C. Schrader, Vice-President, International Contracts, Canadian Superior Oil, Ltd, Three Calgary Place, 355-4th Avenue, S.W., Calgary, Alberta T2P OJ1.

Chile

Jose Cataldo, First Secretary, Chilean Embassy, Javastraat 11, The Hague, The Netherlands.

China

Huaiqing Chen, Embassy of the People's Republic of China, The Hague, The Netherlands.

Shujian Li, Embassy of the People's Republic of China, The Hague, The Netherlands.

Democratic Yemen

Mohamed Rageh, Director, Petroleum Exploration Department, P.M.B. Maala-Aden, P.O. Box No. 5176, Maala-Aden.

Ethiopia

Ato Wodajie Abebe, Project Officer, Ministry of Mines, Energy and Water Resources, P.O. Box 486, Addis Ababa.

Finland

Jouko Leskinen, Assistant Vice-President, Neste Oy, Keilaniemi, 02150, Espoo 15.

France

M. Byramjee, Direction des Affaires Nouvelles, TOTAL Exploration, Paris.

Emmanuel Catta, Economist, Direction des Etudes Economiques, Compagnie Française des Pétroles — TOTAL, Paris.

Gilbert H. Royer, Vice-President for Development, Compagnie Générale de Géophysique, B.P. 56, 6, rue Galvani, 91301 Massy.

Gambia

W. Whyte, Senior Geologist, Ministry of Local Governments and Lands, Banjul.

German Democratic Republic

Volker Mierisch, Oil Production Engineer, VEB Kombinat Erdöl-Erdgas, GDR-3304 Sommern.

Germany, Federal Republic of

F. Bender, President, Federal Institute of Geosciences and Natural Resources, Postfach 51 01 53, D-3000 Hannover 51.

Klaus Erich Koch, Director, Chief Subdivision, Economic Geology, Federal Institute of Geosciences and Natural Resources, Postfach 51 01 53, D-3000 Hannover 51.

Guyana

Clemsford A. Pollydore, Petroleum Engineer, Ministry of Energy and Natural Resources, 41 Brickham and Boyle Place, Georgetown.

Hungary

Lezzlo Modranszky, Director of Commerce, Chemo komplex, 1062 Budapest, Nepkostarsasay utja 60.

Karoly Molnar, General Manager, Geophysical Exploration Company, Geofizikai Kutato Vallalat, 1068 Budapest, Gorkij fs. 42.

Attila Somfai, Chief Geologist and Deputy Manager, Petroleum Prospecting Enterprise, Koolajkutato Vallalat, 5001 Szolnok, Munkasor u. 43.

Iceland

Arni Arnason, Deputy Secretary-General, Ministry of Industry, 101 Reykjavik.

Vilhjalmur Luoviksson, Director, National Research Council of Iceland, Langavegur 13, 101 Reykjavik.

Guömunder Palmason, Head of Geothermal Department, National Energy Authority, Grensasvegi 9, 108 Reykjavik.

Indonesia

Zuhdi Pane, Exploration Manager, Coordination Administration for Foreign Contractors, Pertamina, Jl. Merdeka Timur no. 1a, Jakarta.

Moh. Ashari Warga Dalem, Head, Corporate Planning and Organization, Pertamina, Jl. Merdeka Timur no. 1a, Jakarta.

Luki Witoelar, Head of Geological Evaluation and Development, E.P. Research and Development Division, Pertamina, Jl. Merdeka Timur no. 1a, Jakarta.

Iran

Houshang Ohadi, Senior International Oil Affairs Expert, National Iranian Oil Co., Taleghani Ave. 1, Tehran.

Asud Rouhani, Geologist, Exploration and Production Division, National Iranian Oil Co., Taleghani Ave. 1, Tehran.

Farhang Sajjadi, Manager, Resource Exploitation and Utilization Studies, Corporate Planning, National Iranian Oil Co., Taleghani Ave. 1, Tehran.

Iraq

Abdul-Wahab Sheikh Kader, Head, Exploration and Drilling Department, Ministry of Oil, Mansour, Baghdad.

Italy

Gaetano Lancia, Technical Expert, Planning Department, ENI, P. le Enrico Mattei 1, 00144 Roma.

Ernesto Nathan, Vice-President, Oil Investment Planning, AGIP, P. le Enrico Mattei 1, 00144 Roma.

Bruno Velani, Planning Department, AGIP - ENI, P. le Enrico Mattei 1, 00144 Roma.

Jamaica

William Saunders, Managing Director, Petroleum Corporation of Jamaica, Hotel Inter-Continental, 2 King Street (P.O. Box 579), Kingston, 10.

Jordan

Ahmad Dahkgan, Vice-President, Natural Resources Authority, P.O. Box 7, Amman.

Ahmad Rusan, Chief of Division, Senior Engineer, Natural Resources Authority, P.O. Box 7, Amman.

Kuwait

Nawal Al-Rushaid, Controller of Geology and Geophysics, Ministry of Oil, Technical Affairs, P.O. Box 22795.

Abdullah Baroun, General Superintendent, Exploration Department, Kuwait Oil Company (KSC), Ahmadi.

Abbas Naqi, Controller, International Organizations, Ministry of Oil, Economic Affairs, P.O. Box 5077, Safat.

Malawi

Alfred M'Ndala, Senior Geologist, Geological Survey Department, P.O. Box 27, Zomba.

Mauritania

Mohamed Limine Ould Benahi, Chef de Service, Etude et Programme, Direction des Mines, Direction Mines et Géologie, B.P. 199, Nouakchott.

Nepal

J.M. Tater, Deputy Director-General, Department of Mines and Geology, Lainchouv, Kathmandu.

The Netherlands

B. Backer, Ministry of Foreign Affairs, The Hague.

M. Bark, Ministry of Foreign Affairs, The Hague.

M. Berteling, Ministry of Foreign Affairs, The Hague.

Jan Willem Biekart, Student, University of Utrecht, Budapestlaan 4, Utrecht.

A. Borlsma, Ministry of Foreign Affairs, The Hague.

Hugo de Boorder, Senior Lecturer, University of Utrecht, Budapestlaan 4, Utrecht.

H. Dekker, Rietzangerlaan 19, 2566 ST The Hague.

W.M. Floor, Ministry of Foreign Affairs, The Hague.

Robert Hartmans, Student, University of Utrecht, Budapestlaan 4, Utrecht.

C. Jacobs, Ministry of Foreign Affairs, The Hague.

Torben Johansen, Business Development, Executing Protech International, Acaciastraat 72, The Hague.

M.A.F.J. Keulers, Chief, Concessions, Environment and Planning, Ministry of Economic Affairs, The Hague.

T. Malefason, Deputy Head of Section, International Oil Affairs, Ministry of Economic Affairs, The Hague.

R. Murris, Head, Geological Planning, Shell International Petroleum Co., The Hague.

N. Noseir, Engineer/Marketing Support, Thomassen Holland BV, P.O. Box 70, 6990 AB Rheden.

Maarten Pieters, Student, University of Utrecht, Budapestlaan 4, Utrecht.

B. Racke, Ministry of Foreign Affairs, The Hague.

Jan Rouweler, Scientific Assistant, 5301 SP Zaltbommel.

G.A.N.H. Schreuder, Head, Central Marketing Department, Royal BosKalis Westminster NV, P.O. Box 43, Papendrecht.

Siebe Schuur, Ministry of Economic Affairs, Bezuidenhoutseweg 30, The Hague.

P. Schwegler, Head of Section, International Oil Affairs, Ministry of Economic Affairs, The Hague.

R. Steup, Senior Principal, Ministry of Economic Affairs, The Hague.

P. van Buuren, Deputy Director-General for International Co-operation, Ministry of Foreign Affairs, The Hague.

W. van der Linden, Senior Research Associate, Institute for Earth Sciences, University of Utrecht, Budapestlaan 4, Utrecht.

J. van der Meer, Adviser, De Groot International Contractors BV, Noordweg 15, P.O. Box 54, 3330 AB Zwijndrecht.

J.A. van Hoeflaken, Exploration Manager, Placid International Oil Ltd, Kon. Julianaplein 15, The Hague.

P. Wirds, Managing Director, De Groot International Contractors BV, Noordweg 15, P.O. Box 54, 3330 AB Zwijndrecht.

New Zealand

Vernon J. Dark, Director, Exploration and Gas Development, Oil and Gas Division, Ministry of Energy, Wellington.

Nigeria

G.A. Adams, General Manager, Exploration and Exploitation Division, Nigerian National Petroleum Corporation, Lagos.

G.A. Adesemowo, Deputy Manager, Joint Ventures Department, Nigerian National Petroleum Corporation, Lagos.

Odoliyi Lolomari, Managing Director, Nigerian Petroleum Corporation, Lagos.

B.A. Osuno, Head, Petroleum Inspectorate, Nigerian National Petroleum Corporation, Lagos.

Norway

Tore I. Sandvold, Deputy Director-General, Petroleum Negotiations Department, Ministry of Petroleum and Energy, P.O. Box 8148 Dep, Oslo 1.

Jon T. Strømberg, Executive Officer, Petroleum Development Department, Ministry of Petroleum and Energy, P.O. Box 8148 Dep, Oslo 1.

Peru

Oswaldo Zevallos Nuñez, Gerente del area exploración-producción, Petróleos del Peru, Paseo de la República 3361, San Isidro, Lima.

Qatar

Mohamed Tawfiq Al-Murshid, Chief, Geology Section, Qatar General Petroleum Corporation, Doha.

Saudi Arabia

Hamed Abdul Hamid Bedaiwi, Senior Geologist, Ministry of Petroleum, P.O. Box 141, Damman.

Abdullah Saleh Dakhil, Director of Administration, Ministry of Petroleum, P.O. Box 141, Damman.

Sudan

Abdel Gadir El Shafie, Director of Petroleum Administration, Ministry of Energy and Mining, Khartoum.

Suriname

Sirahmpersad E. Jharap, Chairman, Government Oil Commission, Ministry of Development, Paramaribo.

Bernhard Halfhide, Legal Adviser, Ministry of Development, Laan van Louiseshof 55A, Kwatta.

Sweden

Göran Norell, New Ventures Manager, Svenska Petroleum Exploration AB, Linnégatan 5, S-114 47 Stockholm.

Syria

A. Nader Nabulsi, Vice-Minister of Petroleum and Mineral Resources, Ministry of Petroleum and Mineral Resources, Adawi Street, Khateeb, Damascus.

Thailand

Sanarm Suensilpong, Senior Geologist, Department of Mineral Resources, Ministry of Industry, Rama VI Road, Bangkok 4.

Uganda

Obetia Olimani, Assistant Commissioner, Geological Survey and Mines Department, P.O. Box 9, Entebbe.

United Kingdom

John Brooks, Head of Exploration and Appraisal Branch, Petroleum Engineering Division, Department of Energy, Thames House South, Millbank, London SW1P 4QJ.

Desmond Piggott, Assistant to Director, Exploration and Production, British Petroleum Co. Ltd, Britannia House, Moore Lane, London EC2Y 9BU.

William J. Reeve, Consultant, Petroleum Studies Ltd, Swiss Centre, 10 Wardour Street, London W1.

United States of America

Gordon Barrows, President, International Petroleum Institute, 9 East 53rd Street, New York, N.Y. 10022.

Joy Dunkerley, Senior Fellow, Resources for the Future, 1755 Massachusetts Ave., N.W., Washington, D.C. 20036.

Edward Malonis, Senior Co-ordinator, International Government Affairs Department, Conoco Inc., High Ridge Park, Stamford, Connecticut 06904.

Jacques Marchall, Manager, Evaluation and Analysis, Exploration and Producing Division, Mobil Oil Corporation, 150 East 42nd Street, New York, N.Y. 10017.

Frank Mefferd, Senior Counsel, Esso Exploration Inc., P.O. Box 146, Houston, Texas 77001.

Edgar R. Nelson, Vice-President and Director, Texaco International Petroleum Co., Manager — Contracts and Negotiations, 2000 Westchester Avenue, White Plains, N.Y. 10650.

Charles Ries, International Economist, Office of Energy Consuming Countries, Bureau of Economic and Business Affairs, Department of State, 2201 C Street, N.W., Washington, D.C. 20520.

Franklin Tugwell, International Development Co-operation Agency, Department of State, Washington, D.C. 20523.

Uruguay

Jorge Borad, President, Administración Nacional de Combustibles, Alcohol y Portland, ANCAP, Ave. del Libertador Lavalleja, Montevideo.

Carlos Vanrell, Asesor de Directorio, Administración Nacional de Combustibles, Alcohol y Portland, Ave. del Libertador Lavalleja, Montevideo.

Venezuela

Alex Lorenz, Exploration Director, Corpoven, S.A., Torre Oeste, Penthouse, Avenida Libertador, La Campiña, Apdo. 61373, Caracas 1060A.

Yugoslavia

Srboljub Stankovic, Exploration Director, Oour 'Teknika', Nafta-Gas, Novi Sad.

Zambia

Ernest Jere, Senior Lecturer, Department of Metallurgy, and Dean, School of Mines, University of Zambia, P.O. Box 32379, Lusaka.

Nick Money, Assistant Director, Geological Survey, Acting Director, Minerals and Materials Commission, Geological Survey Department and Minerals and Materials Commission, P.O. Box 50135, Lusaka.

John Bwalya Mwalula, Senior Lecturer and Head of Metallurgy Department, School of Mines, University of Zambia, P.O. Box 32379, Lusaka.

United Nations Secretariat

Bi Jilong, Representative of the Secretary-General, Under-Secretary-General, Department of Technical Co-operation for Development, New York, N.Y. 10017.

United Nations Conference on Trade and Development (UNCTAD)

M. Nagao, Economic Affairs Officer, Technology Division, UNCTAD, Palais des Nations, Geneva, Switzerland.

United Nations Industrial Development Organization (UNIDO)

Issam El-Zaim, Special Advisory Group on Energy (SAGE), Office of the Executive Director, UNIDO, P.O. Box 300, Vienna 1400, Austria.

United Nations body:

United Nations Development Programme (UNDP)

Zsuzsa Bekker, External Relations Officer, UNDP European Office, Palais des Nations, Geneva, Switzerland.

Specialized agency:

World Bank

Philippe Bourcier, Assistant Director, Energy Department, World Bank, 1818 H Street, N.W., Washington, D.C. 20433.

Thomas Fitzgerald, Senior Geologist, Energy Department, World Bank, 1818 H Street, N.W., Washington, D.C. 20433.

Jean-Louis Ginnsz, World Bank, European Office, 66 Avenue d'Iéna, 75116 Paris, France.

Charles McPherson, Contract Specialist, Energy Department, World Bank, 1818 H Street, N.W., Washington, D.C. 20433.

Charles Vuylsteke, Legal Department, World Bank, 1818 H Street, N.W., Washington, D.C. 20433.

Other intergovernmental organizations:

Commission of the European Communities (CEC)

Enro Millich, Oil Supply Division, Directorate of Hydrocarbons, Directorate-General for Energy, Commission of the European Communities, Rue de la Loi 200, B-1049 Brussels, Belgium.

Energy Organization of Latin America (OLADE)

Luiz Augosto Fonseca, OLADE, Casilla 119-A, Quito, Ecuador.

International Energy Agency (IEA)

Jaroslaw R. Martyniuk, Principal Administrator, Oil Industry Division, International Energy Agency, 2 rue André Pascal, 75775 Paris CEDEX 16, France.

Organization of Arab Petroleum Exporting Countries (OAPEC)

Samih Masoud, Senior Economist, OAPEC, P.O. Box 20501, Kuwait.
Abdullatif Y. Zarrug, Senior Petroleum Expert, OAPEC, P.O. Box 20501, Kuwait.

Organization of Petroleum Exporting Countries (OPEC)

R. Sadare, Petroleum Geologist, Energy Studies Department, OPEC, Obere Donaustrasse 93, 1020 Vienna II, Austria.

Secretariat of the Meeting

Vladimir Baum, Director of the Meeting, Director, Natural Resources and Energy Division, Department of Technical Co-operation for Development, United Nations, New York, N.Y. 10017.

C. Constantinou, Technical Secretary and Rapporteur, Chief, Conventional Energy Section, Energy Branch, Natural Resources and Energy Division, Department of Technical Co-operation for Development, United Nations, New York, N.Y. 10017.

Christina Freedman, Conference Officer, United Nations, New York, N.Y. 10017.

Hildegard Eder, Administrative Assistant, United Nations, New York, N.Y. 10017.

LIST OF CONTRIBUTORS AND PANELLISTS

D.H.N. Alleyne, UN Petroleum Adviser to the Government of Kenya, c/o Resident Representative UNDP, P.O. Box 30218, Nairobi, Kenya (p. 127).

Philippe Bourcier, Assistant Director, Energy Department, World Bank, 1818 H Street, N.W., Washington, D.C. 20433, USA (p. 193).

Jean Favre, Director, Institut Français du Pétrole, B.P. 311, 92506 Rueil-Malmaison, France (p. 27).

Martin Forrer, Exploration Consultant, 109 Bank Street, New York, N.Y. 10014, USA (p. 165).

Eduardo J. Guzman, Adviser to General Director, Mexican Petroleum Institute, Av. Cien Metros No. 152, Apdo. Postal 14-805, México 14, D.F., Mexico (p. 233).

Michel T. Halbouty, Petroleum Consultant, The Halbouty Center, 5100 Westheimer Road, Houston, Texas 77056, USA (p. 3).

P. Kassler, Head of Exploration/Production Economics, Shell International Petroleum Co., P.O. Box 62, Carel van Bylandtlaan 30, 2501 AN The Hague, The Netherlands (p. 103).

F.M. Le Leuch, Institut Français du Pétrole, B.P. 311, 92506 Rueil-Malmaison, France (p. 27).

Alfred O. Munk, Manager, Foreign Affairs, Standard Oil Company (Indiana), 200 East Randolph Drive, P.O. Box 5910-A, Chicago, Illinois 60680, USA (p. 205).

Peter R. Odell, Director, Centre for International Energy Studies, Erasmus University, P.O. Box 1738, Wytemaweg 2A, Rotterdam, The Netherlands (p. 17).

Francisco R. Parra, Director, International Energy Development Corporation, 18, Rue Le Corbusier, CH 1208 Geneva, Switzerland (p. 177).

Abdul Aziz Showail, General Manager, Oil and Gas Division, PETROMIN, P.O. Box 67, Dhahran Airport, Dhahran, Saudi Arabia (p. 227).

Michael Tanzer, President, Tanzer Economic Associates, Inc., 251 West 86th Street, New York, N.Y. 10024, USA (p. 213).

Arthur Whiteman, Professor, Petroleum Exploration Studies, Marischal College, Aberdeen University, Scotland (p. 51).

Jan-Olaf Willums, Head Corporate Planning, Volvo Energy Corporation, N-1410 Kolbotn, Vestliveien 12, Oslo, Norway (p. 119).

Hasan S. Zakariya, Interregional Adviser, Petroleum Economics and Legislation, Natural Resources and Energy Division, Department of Technical Co-operation for Development, United Nations, New York, N.Y. 10017, USA (p. 243).

Panelists

Piero Bernardini, Director for International Legal Affairs, ENI, P. le Enrico Mattei 1, 00144 Rome, Italy.

Bernard Taverne, Senior Concession Lawyer, Lecturer, Delft Technical University, Shell International Petroleum Maatschappij, P.O. Box 62, Carel van Bylandtlaan 30, 2501 AN The Hague, The Netherlands.